THE
EARTH
A BIOGRAPHY OF LIFE

THE
EARTH
A BIOGRAPHY OF LIFE

THE STORY OF LIFE ON OUR PLANET THROUGH 47 INCREDIBLE CREATURES

Elsa Panciroli

CONTENTS

11,000 years ago
Warm stable climate

2.6 million years ago
**First hominins
(incl *H. sapiens*)
Ice ages**

23 million years ago
**First horses
Cooling of climate;
Polar ice forms; Drop in sea levels;
Land bridges between continents;
First hominids (incl. Lucy)**

66 million years ago
**Global warming (PETM)
Modern mammal
diversification**

145 million years ago
**Heyday of dinosaurs
First flowering plants
Cretaceous Terrestrial Revolution
Modern mammal origins**

201 million years ago
**Flying and marine reptiles
Dinosaurs diversify
First birds
Middle Jurassic diversification**

252 million years ago
**Origin of modern amphibians
Biggest mass extinction event in history
First mammals; Reptiles diversify
Birth of modern ecosystems**

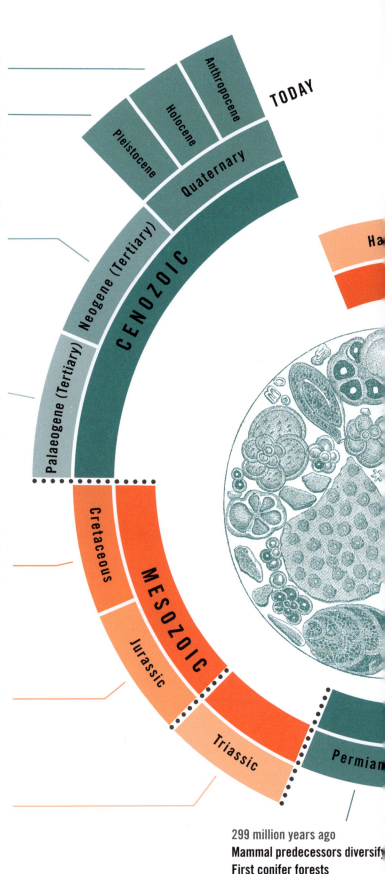

TODAY

Anthropocene

Holocene

Pleistocene

Quaternary

Neogene (Tertiary)

Palaeogene (Tertiary)

CENOZOIC

Ha

Cretaceous

MESOZOIC

Jurassic

Triassic

Permian

299 million years ago
**Mammal predecessors diversify
First conifer forests**

TIMELINE OF LIFE ON EARTH

•••••• **Biggest mass extinctions**

TODAY

CENOZOIC
66 million years
1.5% of geologic time

MESOZOIC
186 million years
4% of geologic time

PALAEOZOIC
289 million years
6.3% of geologic time

RELATIVE LENGTHS OF GEOLOGIC PERIODS

PRECAMBRIAN
4,059 million years
88.2% of geologic time

EARTH FORMS

4,600 million years ago
Earth forms

4,000 million years ago
**Earth's crust forms; Plate-tectonics
begins; Life begins in sea**

2,400 million years ago
Great Oxygenation Event

635 million years ago
First complex life

541 million years ago
**First chordates
(animals with backbones)
Cambrian explosion**

485 million years ago
**Major diversification of animal life
First jawed fish
First land plants**

444 million years ago
**First vascular plants
First arthropods on land**

419 million years ago
**Fish diversity
First vertebrates on land**

359 million years ago
**Trees and seed ferns flourish
Amphibian lineage diverges
Reptiles and mammals diverge
Carboniferous rainforest collapse**

Archaean

PRECAMBRIAN

Proterozoic

Ediacaran

Cambrian

Ordovician

PALAEOZOIC

Silurian

Devonian

Carboniferous

INTRODUCTION

DEEP TIME

Our planet's lifespan is measured on a timescale too vast to comprehend. Centuries of studying rocks and fossils have allowed us to assemble the story of Earth's history and development over millions of years. Humankind has long sought to understand the world by reading the landscape, but only in the last few hundred years have we grasped the awesome ways in which the surface of the planet has been created and re-arranged many times over, and how this shaped the path of life on Earth.

The Earth is made of rocks, from her molten heart to the tough skin of her surface. The study of rocks and fossils (geology and palaeontology) has given humankind an unparalleled glimpse into our planet's formation, development and antiquity. Despite the information we've gleaned, however, comprehending geological time – also known as deep time – is difficult, as it reaches far beyond the realms of human experience.

Geologists study the composition, age and distribution of rocks. This has illuminated processes such as plate tectonics, climate change and the origins of life and evolution. The principles of geology are both deceptively simple and sneakily complex. In deep time, solid rock can flow like water and crumple like paper. New rocks form while others are swallowed again. Meanwhile, fossils are distributed unevenly not only in terms of where they are located, but also which particular slices of time are represented. Animals with skeletons fossilize more readily than those without. Putting this puzzle together began in earnest around three hundred years ago, but humankind has been trying to make sense of the world for much longer, from deciphering shells on mountaintops as evidence of ancient floods, to using mythology to interpret dinosaurian bones emerging from the desert sands of Mongolia's Silk Road.

Geologic Timescale

The geological timescale is used to chart the timing of events through the Earth's 4.6 billion-year lifespan. It is split into increasingly fine timescales, from eons to eras, to periods then ages. Most of these were named by European geologists and defined based on obvious changes they saw in layers of rock, such as a sudden transition from limestone to sandstone, or the appearance of new types of fossil not seen in older layers. As our understanding of the processes of geology has improved, the details of these time periods have been refined, and the dates are increasingly precise.

The modern timescale used by scientists is called the chronostratigraphic chart, and it combines information from multiple sources, including the dating of radioactive elements in rocks and fossils. Yet, despite all of the advances in technology, one of the key principles underlying the geologic timescale is still the study of fossil organisms, and how they appear, change and disappear through deep time.

Stories in the Rock

The first recorded discourse about rocks comes from Ancient Greece and Rome, where authors examined stones, metals and minerals, and recognized that the Earth had changed a great deal over time. Around 1000 CE, Persian and Chinese thinkers were using the composition of rocks to hypothesize about how landscapes were formed. Ibn Sina (981–1037), one of the greatest polymaths of the Islamic world, recognized the incredible timescales necessary for rocks to be laid down and valleys gouged out. In China, Shen Kuo (1031–95) likewise saw these processes of erosion and deposition, and that the fossils of shellfish showed that parts of inland China must once have been under the sea.

One of the main figures in Western geological science is James Hutton (1726–97). Often dubbed the father of geology, he made the same observations about erosion and deposition as the great thinkers before him, realizing that these processes in the present day revealed the formation of rocks in the past. Hutton formulated new ideas about deep time, as well as recognizing that rock could be lifted, tilted and folded to create mountains, valleys and perplexing formations. This was the beginning of our modern understanding of geological processes.

Made of Many Layers

There are three main rock types on Earth: igneous, sedimentary and metamorphic. Igneous rocks come from below the Earth's surface, being either spewed out in eruptions as lava, or solidifying beneath the surface from magma. Sedimentary rocks, on the other hand, accumulate on the surface of the planet, composed of pieces of eroded rocks and minerals, fossil organisms or chemical precipitates (such as carbonates). Metamorphic rocks begin life as igneous or sedimentary rocks, but are changed by being compressed or heated. Rock can be metamorphosed when it comes into direct contact with the intense heat of magma, or it can be folded, pressed and squeezed like putty. This often alters the chemistry of the rock,

creating new textures and patterns as the minerals re-arrange themselves.

The most important principles in geology are tied intimately to understanding their great age. Rock layers lie one on top of the other, like tiers in a cake; the oldest layers on the bottom, becoming younger as you move to the surface. Fossils in the layers show changes in organisms over time, and index fossils – species that exist only during distinct time periods – can be used to date rocks. However, in deep time, layers can be tilted or folded, pushing older ones over younger. An example of this is the Moine Thrust along the northwest coast of Scotland. Layers can be also scoured away by rain and ice, creating gaps in the rock record. Volcanic activity shoots magma through existing rock layers like lightning bolts. Cracks in the Earth's surface, fault lines and plate tectonics shift layers relative to one another, creating complex and confusing patterns. Interpreting these patterns is challenging, and they can only be understood when the vastness of the Earth's lifespan is taken into account.

Restless Plates

Continental drift is part of the mechanism of plate tectonics, one of the most fundamental processes that shapes our planet. Although it appears as a solid coating, the Earth's surface actually comprises plates of rock. There are around eight main plates and dozens of smaller ones, and they shift restlessly due to the boiling of the Earth's molten core, churning beneath them. This heat creates convection currents that carry the plates, grinding them together and pulling them apart over millions of years. Depending on their thickness, plates slide under one another or fold and crush upwards to create mountain chains. Where plates meet and diverge, volcanos and earthquakes are frequent, as at the Ring of Fire around the edges of the Pacific.

Many of the larger plates contain an ancient core called a craton. These are the oldest pieces of the crust, some of them having formed over four billion years ago, shortly after the birth of the planet. By studying cratons, geologists have been able to piece together Earth's formation. The shuffling of continents over the last 3.5 billion years – and particularly since the emergence of complex life – has had a heavy hand in the process of evolution, creating new habitats, opening and closing seas, redirecting the climate and isolating organisms from one another for millions of years.

Patterns of Life

Evolution shines the light by which we can see the technicolour history of our planet. Recognizing its processes began relatively recently but has completely transformed the study of biology and fossils. The link between evolution and the physical environment is intimate, tying the patterns of life to the ever-changing Earth.

The process of evolution was famously recognized first by Charles Darwin (1809–82), and later by Alfred Russel Wallace (1823–1913). Evolution by natural selection fundamentally shapes how we interpret the patterns of life on Earth, both now and in deep time. Studying evolution combines biology, palaeontology, geology, ecology and mathematics. It is a seemingly simple concept – the inheritance of characteristics and how they relate to survival – but encompasses such intricate complexity that it is easily misunderstood and mischaracterized.

With the advent of computers and genetics, our knowledge of how traits are selected and passed down through the generations is more refined than ever. Fundamental to that knowledge is the study of rocks and fossils, which provide information about change on timescales unobservable even over multiple human lifespans. We can appreciate how the changing face of the Earth has shaped her inhabitants. Without this insight into the past, it would be impossible to understand how our present world was formed, or what the future may hold in the face of climate change.

Hopeful Monsters, Common Ancestors

The fundamental lynchpin of evolution is that all life on Earth evolves from common ancestry. Descendants inherit traits from their ancestors, and traits that provide advantages for survival spread through the population over generations. Although this seems straightforward, for a long time people saw selection as an active process striving to perfection or improvement – an idea that persists today. Evolution has no end-goal, and traits are not actively selected or developed by an organism but instead constitute a biological lottery for every generation. Evolution is an ongoing process with no value judgements.

Animals have been arranged into groups since Carl Linnaeus (1707–78) created his system of taxonomy in the 1700s. This was based on anatomy, linking animals (including fossils) together through shared traits in their skeletons and organs. Modern science uses cladistics rather than Linnaean taxonomy. Cladistics doesn't rank organisms but instead incorporates anatomy and genetics to group organisms based on common ancestry. These groups are called clades, and reflect a better understanding of the true relationships between organisms, and the process of evolution.

In the last few decades, genetics has rewritten animal relationships rendering many original classifications obsolete. Genetics has also exposed how the separation of animals by plate tectonics has led to them independently flourishing on each continent – often evolving similar adaptations for survival through a process called convergent evolution. Yet we have access only to the genes of *living* animals; fossils and the study of their anatomy still play a crucial role in interpreting the bigger picture of the evolution of life in deep time.

Tempo and Mode

Knowledge of genetics, the application of mathematics and developments in computing over the last century have led to a revolution in biological science, called the modern synthesis. This began in the first half of the 20th century. Whereas simple observation – which could be influenced by the skill or assumptions of the observer – had been the sole method for understanding animal relationships and natural selection before, these new approaches were quantitative and could be tested mathematically.

Fossils provide crucial data to understand the speed at which evolution takes place and the major patterns it has followed. They show that evolution can occur slowly and incrementally, as Darwin predicted, but that at times it can also be rapid. There have been bursts where animals have radiated out into many different new species. Evolution is not linear and directional but irregular, with many branches and no set goal. Using

mathematics, it is possible to match up changes in the course of evolution with major events such as extinctions and climate change. We can conceive a more complex picture of natural selection, and this has transformed how we see the history of life on Earth.

Ecosystems and Life

A gloriously messy web of life and death transfers energy through our planet's ecosystems. An ecosystem encompasses all the plants and animals – including micro-organisms – and their intimate interactions with geology, climate and one another. Ecosystems can be understood through the flow of energy and materials, which cycle in a pass-the-parcel game through ecosystems via photosynthesis, predation, decomposition and nutrient recycling. These interweaving webs of energy first emerged as life began, and as organisms have grown increasingly complex so their interactions have followed suit. Evolution is wound together tightly with ecosystem change.

Climate change and natural disasters have also shaped ecosystems in deep time, disassembling the intricate webs of life that inhabit them. This destruction shuffles the deck of life, giving selective advantages to new groups of animals. At points in the history of life on Earth, entire ecosystems have collapsed, carrying their inhabitants with them into the fossil record. There have also been momentous 'births' of new types of ecosystem, underpinned by novel energy sources or specific predator–prey relationships.

PRECAMBRIAN

Around 88 per cent of our planet's lifespan took place in the Precambrian. From the clumping of dust and space rocks to create the first foetal globe 4.6 billion years ago, to the emergence of complex life in the ocean 4 billion years later, the Precambrian spans not only the birth and childhood of our planet, but sees Earth well into maturity.

The Precambrian is an informal name for a division of time comprising the Hadean, Archaean and Proterozoic Eons. It is so named because it lies before the Cambrian, once thought to be the dawn of life. We now know that cellular life sparked into being long before, perhaps within the first billion years after Earth was formed, and that multicellular lifeforms were making the seabed their home before the Precambrian had drawn to a close.

Despite the aching chasm of time represented by these eons, we know comparatively little about the Precambrian. The rocks formed in these early days have mostly been thriftily recycled, and those that remain are gnarled to near oblivion. Yet, enough remain to illuminate the fundamental steps our planet underwent to become habitable for the complex animals that now riddle every nook of its surface.

The Precambrian began in the Hadean with an oxygenless cauldron planet, zapped by the sun's radiation and battered by asteroids. The Earth collided with another protoplanet, and our moon was formed from their union. Oceans were poured and lost again, until finally the searing surface of the planet cooled enough to retain its moisture. Some studies suggest that shortly after the oceans became permanent, the first single-celled life may have sparked into being somewhere in the depths.

In the Archaean the seas were hot and green as pea soup. An atmosphere had formed, but it was a miasma of toxic gases. Yet, this was when single-celled life multiplied. Soon the magical process of photosynthesis began; absorbing energy from sunlight and sputtering out oxygen as a by-product. Single-celled sun-eaters were steadily shifting the ratio of atmospheric gases, unknowingly creating a planet fit for complex life.

The first continents formed in the Archaean, and by the Proterozoic plate-tectonics had formed. Supercontinental cycles bunched and split the crust over millions of years, drawing up mountains and then rubbing them out again. Increasing oxygen ousted the previously abundant greenhouse gas, carbon dioxide, leading to a global freeze called Snowball Earth. This tough environment may have helped trigger the rise of the first complex organisms. Shortly after the ice melted near the end of the Precambrian, the squishy bodies of bizarre lifeforms are found in rocks from the ocean floor.

Our planet was finally home to complex organisms, multiplying in a thriving sea, under an oxygenated sky. The world would never be the same again.

Hadean

4.6 to 4 billion years ago. To begin with most of the Earth's surface was molten, with surface temperatures over 200 degrees Celsius (390 degrees Fahrenheit). Asteroids frequently hit the planet.

Crust began to form by the end of the Hadean.

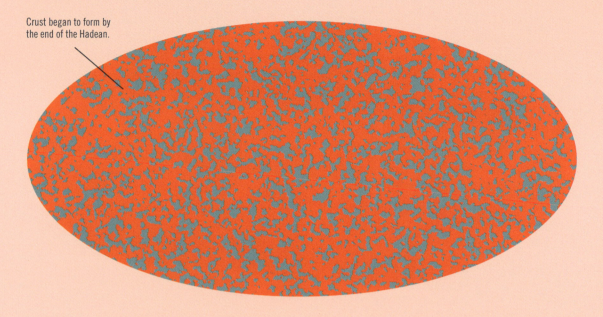

Archaean

4 to 2.5 billion years ago. First life may have formed around deep sea thermal vents.

First landmasses formed – their remnants still survive today.

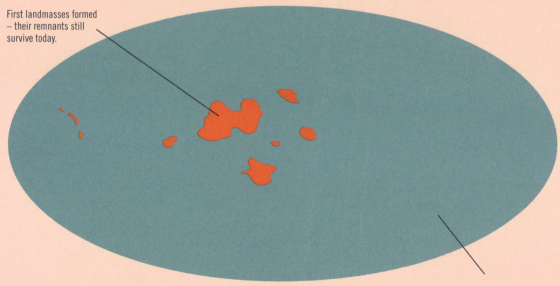

Oceans formed from water brought by asteroids. Water was hot and tinged green by iron ions.

Snowball Earth

Over 635 million years ago, during the Proterozoic (2.5 billion to 541 million years ago). Global ice cover led to an 'albedo effect', reflecting sunlight and further cooling the planet.

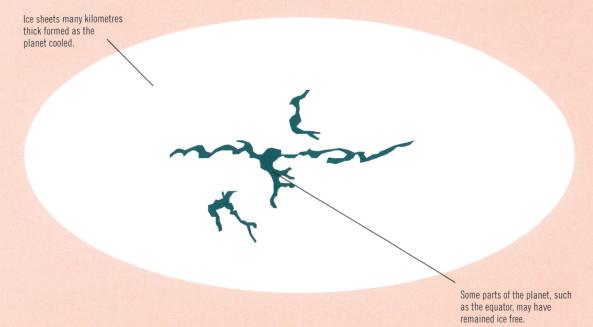

Ice sheets many kilometres thick formed as the planet cooled.

Some parts of the planet, such as the equator, may have remained ice free.

Ediacaran

635 to 541 million years ago. Complex life emerged in the sea, much of it soft-bodied.

The moon was closer, resulting in higher tides on Earth's shorelines.

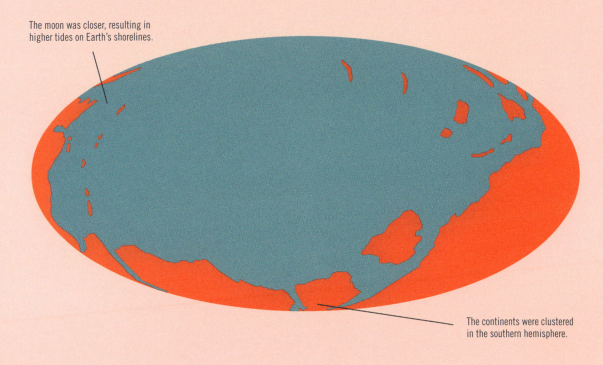

The continents were clustered in the southern hemisphere.

Hadean

The Earth began in the hellish Hadean, which lasted 600 million years. The moon was formed from an impact with another protoplanet, bringing the water that would form our atmosphere and oceans. Although Earth seemed inhospitable, by the end of the Hadean our oldest common ancestor had appeared, and the epic tale of life's evolution began.

In the first days of the Earth our planet was unrecognizable. It formed approximately 4.6 billion years ago from dust grains orbiting the newborn sun. These grains clumped together and collided, growing to form the first protoplanets, including our own. The process only took around twenty million years – a mere blink in geological time. The Earth was a hot mess, battered by a rain of impacts from other protoplanets and asteroids, which kept most of the surface molten. Heavy metals in this magma soup sank to form the core, which produced the Earth's magnetic field. The name Hadean – from the Greek god of the underworld, *Hades* – was chosen for this time period because the extreme conditions were reminiscent of the Western Classical concept of hell. For much of this time period, Earth was incompatible with life: the surface temperature was over 200 degrees Celsius (390 degrees Fahrenheit); there was almost no oxygen; and the lack of ozone layer meant the planet was flooded with deadly radiation emanating from the furious sun.

Slowly, Earth cooled and formed a crust. By the end of the Hadean, this crust began to move, and plate-tectonics was set in motion. A day – one full rotation of the Earth – was only four hours long at first, gradually lengthening to ten hours by the end of this eon. The sun was only 70 per cent as bright as it is today, and the moon was much closer and would have been enormous in the sky. Towards the end of the Hadean, a huge number of asteroids pummelled the Earth and other planets in our young solar system. Evidence for this is found on the pockmarked moon, which bore the brunt of impacts. In the last fifty years, scientists have found Hadean rocks in Australia, Canada and Greenland, but most have been recycled through plate-tectonics.

Although the earliest fossil evidence is not found until much later, researchers believe the first life on Earth may have appeared at the end of the Hadean. They call it LUCA (the Last Common Universal Ancestor), and from it all life on our planet evolved. It probably began through chemical reactions that produced simple organic compounds, including the building blocks of life. LUCA can be studied using the genomes of living species, including single-celled organisms. Scientists trace these patterns back through time using molecular clock analysis, which counts the number of mutations in genes to work out when different branches in the tree of life split. From the myriad protocells that developed on the early planet, only one lineage survived. This simple life diversified, transforming the Earth and creating conditions that made it possible for more complex organisms to emerge.

The Moon is Born

Our planet's moon is thought to have formed when a protoplanet called Theia collided with Earth around 4.4 billion years ago. A huge part of the Earth would have melted, vaporizing rock and existing water to form the first atmosphere as they condensed, leaving behind CO_2, hydrogen and water vapour. A disc of debris from the collision orbited Earth, eventually coalescing to form the moon. It was molten initially, and developed a small iron core before cooling and solidifying in as little as a hundred years. Evidence for the moon's origin comes from moon rocks themselves, which have identical oxygen ratios to those on Earth.

The impact with Theia was perhaps the most important event in our planet's early history. It may have nudged the Earth, causing it to tilt on its axis. This 23-degree tilt produces our seasons, as the high latitudes periodically turn their cheek away from our warming star. Without the moon, we would have no tides. To begin with, it was almost twenty times closer to the Earth than it is now, and has been slowly drifting away into space ever since, at a rate of around 4 centimetres ($1\frac{1}{2}$ inches) per year. This is happening because of the interaction of the moon's

gravity with our oceans, which causes the Earth's spin to slow and the moon's orbit to speed up, flinging it outwards. As the moon departs, our planet's days will lengthen by around 19 hours every 4.5 billion years.

Planet Ocean and Life

Oceans are the cradle of biological life, but they didn't always exist. It isn't clear where the first water came from, because the temperature during the Earth's formation was too hot for it to exist. Water probably arrived on our newly formed planet via impacts with water-rich asteroids and protoplanets, including the impact that formed the moon.

Despite the temperature on the Earth's molten surface, the intense atmospheric pressure prevented the early oceans from evaporating. However, water was lost to space more easily due to the smaller mass of our planet, which meant molecules could escape Earth's gravitational pull. Seas may have formed and been nearly or completely lost more than once in the first few hundred million years. At least one ocean's worth of water was lost in the first billion years.

By 4.4 billion years ago, the Earth had cooled sufficiently for more permanent seas and rainfall. Our oceans now cover 71 per cent of the planet's surface, and water is even incorporated into minerals within the Earth's crust, mantle and core, carried there as continental plates slide under one another. It is estimated up to three times as much water might be stored within the planet as there is on the surface.

Archaean

The Archaean Earth was an ocean otherworld. The seas were hotter than a bath and tinged green. The atmosphere reeked of poisonous gases, and surface temperatures were searing. Microbial life flourished, releasing the first breaths of oxygen, but it would be billions of years before complex organisms arose.

For 1.5 billion years the Earth remained a harsh environment for life. Archaean rock formation dating to between 4 to 2.5 billion years old still exist, but are often heavily metamorphosed; having been heated over the eons and folded repeatedly like towels. The first continents formed in the Archaean, and pieces of them still survive. Plate-tectonics had been set in motion, but the continents themselves were completely different. There were more volcanic eruptions than there are today due to the heat rising from the planet's core, and these created island arcs, which drifted together to form the first landmasses.

The Archaean proto-continents were surrounded by a hot lime-green ocean. The water was rich with iron ions that gave it this alien tinge. At first this covered almost the entire globe, and was as much as 85 degrees Celsius (185 degrees Fahrenheit). It cooled over the next two billion years. Air temperatures were lower than in the preceding Hadean, but the atmosphere was still toxic to most life, being filled with methane, ammonia and CO_2. These gases created a greenhouse effect that baked the globe. A day lasted just 12 hours, and the sun was still only 75 per cent as bright as it is today. Yet this strange planet was

the cradle of life, teaming with micro-organisms. By the end of the Archaean, the first wafts of oxygen were permeating the atmosphere, a precursor to the great oxygenation that would come later in Earth's history.

Life, Not As We Know It

Most fossils of early micro-organisms are found in rocks formed in warm, shallow water environments, but there is other evidence for life in the most unlikely places. Tube-like structures in rocks over 3.7 billion years old may be the remains of micro-organisms living around hot, or hydrothermal, vents on the deep seabed. These vents exist today, often near tectonic plate margins or hotspots in the Earth's crust. They form towers where seawater gushes through volcanically heated rocks. No light reaches the seabed, but chemicals dissolved in the water sustain a small but surprisingly complex food chain of micro-organisms, clams, shrimp and tube worms. This is called chemotrophy, from the Greek meaning 'chemical feeding'.

Photosynthesis Begins

The first photosynthesizing organisms evolved in the Archaean. These were simple prokaryotes: cellular organisms such as bacteria and archaea that don't have an enclosed nucleus within the cell. Eukaryotic cells, on the other hand, which evolved much later, are more complex in structure and have a membrane enclosing their nucleus. All complex animals are eukaryotic, as well as plants, fungi and protists.

The first prokaryotes probably fed on dissolved chemicals, but later some began to use the rays of the sun to fuel their metabolic processes, creating oxygen as a waste product. The cyanobacteria in stromatolites (see below) are thought to have played a major role in pumping our atmosphere with this life-giving gas. It was almost a billion years before oxygen levels rose high enough to fuel the evolution of complex animals.

Thin Layers of Life

Some of the most ancient forms of life on the planet are stromatolites. They resemble lifeless rocks but their surface teems with millions of photosynthetic cyanobacteria, fuelled by the power of the sun. Stromatolites form columns, mounds or cones that can be over a metre (3 feet) across, but it is only the surface that contains the living organisms. The shapes come from the deposition of sediments on the surface, bound together by mucus secreted by the cyanobacteria. This hardens into calcium carbonate, and builds up into layers like onion peel. These layers can be seen in the rock record in places that were once wide, shallow seas, and represent some of the oldest fossil evidence for life on Earth.

Some stromatolite structures can be created through non-biological processes, for example by precipitation of carbonates from seawater. It's difficult to distinguish biological versus non-biological stromatolites in the fossil record, but the structure of some fossil stromatolites are too complicated to have formed without the aid of living organisms. In the Archaean and Proterozoic, stromatolites were common, but as more complex organisms evolved, they grazed on stromatolite colonies and reduced their numbers. They are now rare, existing mostly in extremely salty environments (hypersaline), such as Shark Bay in Australia, which protect them from being grazed by animals.

Proterozoic

The Proterozoic is the longest eon in the Earth's geologic timescale, lasting a staggering two billion years. It holds a tumult of events, from the infusion of oxygen that transformed the course of evolution, to the global glaciations that nearly wiped it out again. Sexual reproduction emerged for the first time, and by the end of this eon, the first multicellular life forms populated the seabed: the beginning of the bizarre and beautiful animal world.

The Proterozoic accounts for over 40 per cent of the whole lifespan of the planet. It began 2.5 billion years ago, at the close of the Archaean, and ended 541 million years ago. During this time, Earth was transformed into a world we recognize as our own. A day reached 23 hours in length, and the atmosphere and oceans accumulated more and more oxygen. Most of the planet remained covered in water: the Panthalassic and Pan-African oceans. The Earth was very tectonically active, with dry land rising from the depths as mountains were pushed skywards and volcanos erupted. Around 43 per cent of the continental crust was formed in the Proterozoic, and this crust was increasingly stable, better able to endure the destructive processes of deep time. The fresh continents would have remained almost barren, with only bacterial and later fungal organisms colonizing the surface.

Great Oxygenation Event

Although oxygenation of our planet began in the Archaean, near the start of the Proterozoic there was a sudden leap in oxygen levels in the atmosphere. Dubbed the Great Oxygenation Event, it is thought to have been caused by new types of cyanobacteria, which released oxygen as a by-product of photosynthesis. At first this oxygen reacted with iron and sank to the sea bottom as particles of rust, but after fifty million years or so, the levels of iron were low enough that the amount of oxygen rose up into the atmosphere instead of being swallowed by the sea.

Oxygen was essential to the emergence of complex life. Multicellular organisms could not have evolved without freely available oxygen to fuel the chemical reactions in their cells. Oxygen is highly reactive and readily forms new compounds, which created ecosystems packed with nutrients. Nitrogen was converted into compounds such as ammonia in a process called nitrogen fixation, making it available for essential biological processes.

There was a dark side to the radical atmospheric change: the rise in oxygen came at the expense of greenhouse gasses. Unlike our current climate change crisis, the lack of carbon dioxide and methane in the Proterozoic caused global temperatures to plummet, leading to worldwide glaciation.

Snowball Earth

Just over 630 million years ago, our planet was encased in ice. This happened more than once, and though it seems counterintuitive, could have helped kick-start the rise of complex life on Earth. The evidence for Snowball Earth phases in the rock record comes from glacial sediments and 'drop stones', which were scraped from the surface of the ancient continents by glaciers and carried out to sea, where they sank. These are found in rocks from this time in almost every part of the Earth, including what was then the tropics.

There are many possible causes for Snowball Earth: greenhouse gases in the atmosphere being replaced by oxygen, changes in the Earth's orbit and sun's solar energy output, and volcanic eruptions have all been blamed. The position and break-up of the supercontinent of Rodinia may have altered ocean chemistry and contributed to thickening ice cover. The growing glaciers created what is known as an albedo effect, where the reflective white surface sends the sun's rays back into space, cooling the planet further. Global temperatures plummeted lower than the modern-day Antarctic. Beneath the frigid ice, life clung on.

Over millions of years, volcanic eruptions and respiring microbes released enough CO_2 and methane to raise temperatures and melt ice over large parts of the ocean. A layer of ocean slush up to a mile (2 kilometres) thick formed. Uncovered areas of dark water reversed the albedo effect and absorbed the sun's heat, until eventually the planet was ice-free once more. Organisms had to be tough to endure these radical temperature swings – adapting to this rollercoaster may even have driven their complexity. When the surface defrosted, the scene was set for life to colonize the world.

Eukaryotes – Complex Life Begins

Eukaryotic cells are the basis of all complex, multicellular life on Earth. They emerged as much as 2.7 billion years ago through the joining of two independent cells. Eukaryotes invented sexual reproduction, and with it came the major mechanism for the process of natural selection.

Eukaryotes are organisms whose cells have a nucleus enclosed in a membrane. They also contain organelles inside them, the parts of the cell that carry out specialized functions, like microscopic versions of the organs in a body. These organelles include mitochondria and chloroplasts. All complex life on Earth today is made of eukaryotic cells, including humans. As the building blocks of every organism on the planet, they hold within them the blueprints for evolution.

The first definitive eukaryote fossils are found in the Proterozoic. They were a minor component of life on Earth for around a billion years, before erupting into a complex range of life forms. It's difficult to know how they evolved, but the most likely explanation is that two independent, simpler cells formed a mutually beneficial relationship (called symbiosis), and eventually one cell was incorporated into the other, becoming part of it. This is probably how the mitochondria in our cells originated.

Taking a Lichen

Some of the earliest eukaryotes are algae and fungi. Their fossils have been found in rocks around a billion years old, but they probably originated much earlier. Older structures that could be algae or fungi fossils have been identified, and ancient oil found in rocks from 2.7 billion years ago could have formed from simple eukaryotes living as far back as the Archaean. Fungi and algae together form the symbiotic organism known as lichen, which remains common today. In a lichen, either an algae or cyanobacteria live within the fungal filaments, providing their host with nutrients from photosynthesis in exchange for shelter and moisture. The oldest fossil lichen comes from the Rhynie Chert in Scotland and is around 410 million years old.

Another of the earliest eukaryotes is *Caveasphaera*, a 609 million-year-old fossil from China. It is less than half a millimetre in diameter and looks like a tiny ping-pong ball. It probably represents the embryo of a multicellular organism, and may have belonged to a precursor of the emerging animals of the Ediacaran Biota.

Dinoflagellates, single-celled eukaryotic organisms that live in marine and freshwater.

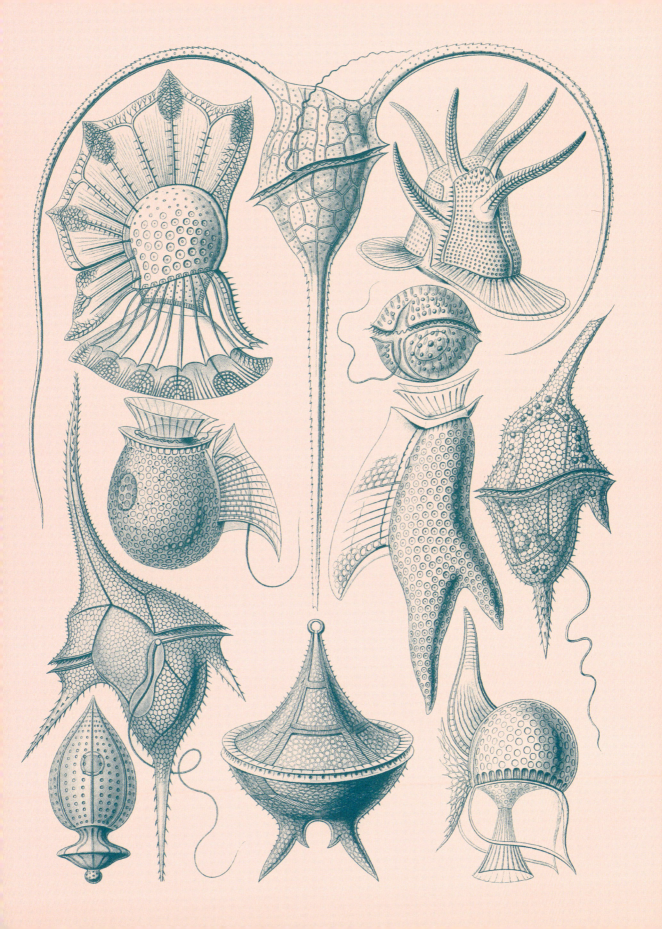

Importance of Sex

Sex has its advantages. It allows genes to recombine, producing random mutations like tickets in the lottery of natural selection. This makes sexually reproducing organisms better able to adapt to changes in the environment, predation pressure, disease and parasites.

Eukaryotic cells reproduce sexually, with two cells each contributing half of the genetic material to their 'offspring'. Sexual reproduction is thought to be a very ancient eukaryotic trick. The first evidence for sexual reproduction comes from a red algae called *Bangiomorpha*, which is just over a billion years old. Sex is fundamental to the process of evolution. Simply put, without sex, life on Earth today would be neither as complex nor as exciting.

Diatoms, a group of algae found in our marine and fresh waters, and in soil.

Ediacaran

In the mysterious Ediacaran, the last time period of the Proterozoic, the first multicelled animals appeared. Not only did unique life forms pepper the ocean floor, but they then also died in one of the first major mass extinctions in evolutionary history. Life at this time is only now being understood, thanks to the latest fossil discoveries.

The elusive Ediacaran lasted from 635 to 541 million years ago. It takes its name from the Ediacara Hills in Australia, where geologists first identified fossils from this time period, and changed our picture of the emergence of multicellular organisms on the planet. Although the Ediacaran Earth was similar to today, the continents were clustered in the southern hemisphere, and the north was one unbroken ocean. The moon was closer to the Earth, driving stronger tides that washed high upon the barren shorelines. By the end of the period, Gondwana had formed, the landmass comprising the core of the present-day southern continents. It would endure for the next 350 million years, being incorporated into the supercontinent of Pangaea before finally breaking apart in the Jurassic.

The first complex multicellular organisms are known as the Ediacaran Biota. These fossils are rare and hard to study because they were mostly soft-bodied organisms, lacking hard mineral skeletons. Their ghostly outlines and intricate impressions include creatures shaped like fern fronds and jelly-like blobs, ranging from 1 centimetre to over 2 metres (2/5 inch to over 6½ feet) in length. Strange and hard to interpret, some of them may represent the ancestors of lineages that survive to the present day.

Avalon Explosion

It was previously thought that complex animal life began in the Cambrian, which comes after the Ediacaran. We now recognize that there was a much earlier radiation of organisms that preceded this in the Ediacaran. They appeared not long after the last phase of global glaciation (Snowball Earth). The sudden increase in fossils at this time is dubbed the 'Avalon Explosion', taking its name from the Avalon Peninsula in Newfoundland in Canada, where exceptionally preserved Ediacaran fossils have been discovered. There are now well-known Ediacaran fossil sites in Australia, Namibia, Russia and China.

The Avalon Explosion saw an increase in the diversity of these pioneer organisms. Over a hundred different types of Ediacaran animal have been described, mostly soft-bodied, although some like *Cloudina* produced a hardened skeletal casing. Many had bizarre bodies unlike later animals, making their evolutionary relationships difficult to figure out.

First Mass Extinction?

The Ediacaran ended 541 million years ago, and most of these unique first animals disappeared. This might be the world's first mass extinction, perhaps caused by a sudden change in ocean circulation linked to the break-up of the supercontinent Rodinia, which may have reduced oxygen levels in the ocean. This is called an anoxic event, and is captured in rock deposits of black shale, which forms in oxygen-poor conditions. If true, this was one of the worst anoxic events in the history of complex life.

Another possible reason for the loss of the Ediacaran Biota is that new types of organism evolved that fed on the microbial mats that bound the surface of the seabed together. These mats were a vital part of the Ediacaran ecosystem, providing stable habitats and anchorage, but they also prevented nutrients and oxygen from circulating between the water and ocean-floor sediments, keeping them barren. In the Earliest Cambrian there is a sudden increase in the number of burrowing animals, which may have broken up the microbial mats, replacing them with a flourishing new Cambrian ecosystem.

Charnia — First Animals

Charnia is a strange animal that lived over 550 million years ago. It is among the first complex organisms to evolve, belonging to an ecosystem scientists still struggle to understand. With a body unlike anything alive today, *Charnia* is one of many bizarre life forms in the Ediacaran. Alongside it dwelled animals with the first mineralized skeletons: armoured hulls to protect themselves against an increasingly hostile world.

Charnia is a frond-like organism that lived in the Ediacaran. It is known from sites in the British Isles, Australia, Russia and Canada. Growing over half a metre (20 inches) long, it resembles a plant, but features of its anatomy tell us that it doesn't belong to this group. It was originally identified as a type of algae, and later a sea pen (a marine animal related to jellyfish). Although they look similar, further studies show *Charnia* grew in a different way, adding new buds on the tip of its fronds rather than at the base. It lived relatively deep on the seabed, where it was anchored with a round hold-fast. It couldn't have photosynthesized due to lack of sunlight, but it didn't have a mouth or gut either. The fronds might have been used to filter-feed or absorb nutrients from the surrounding water. It has a body plan of alternating branches, lacking either the bilateral (down the centre) or the radial (in a circle) symmetry seen in most organisms alive today. Some researchers suggest it was a completely unique organism, possibly with no close relationship to any existing animal groups.

The fossil *Charnia* has become an icon of the Ediacaran Biota. It is the first fossil identified as pre-dating the Cambrian – once thought to be the earliest time period with complex animals. It was discovered in 1956 in Charnwood Forest near Leicester, England. It was originally found by a teenager named Tina Negus, and although she told her geology teacher about the specimen, it was in rocks considered too old to contain fossils, so they discounted her discovery. The following year, a schoolboy called Roger Mason came across the same fossil, and his find was taken seriously. The fossil was subsequently described and named after him, *Charnia masoni*. Later, Negus's role was recognized, and they are now jointly honoured for the discovery of *Charnia*, with Negus being the true first discoverer. Like many creatures from the Ediacaran, there is much about *Charnia* that remains a mystery.

Charnia are one of the oldest complex organisms known.

**Animal,
Vegetable,
Mineral**

————————

Among the strange and squishy animals living on our planet in the Ediacaran, there were some that had developed the first hard mineralized skeletons. *Cloudina* was one of these: an odd stack of fairy-sized cups that often formed 'reefs' of multiple individuals. They grew from the width of a finger to longer than a human hand. Their appearance in life is unknown; it's possible there were soft body parts within or surrounding the animal. Although *Cloudina* is abundant in some rock layers, they are never found alongside their soft-bodied counterparts, suggesting they lived in quite different environments.

The hard skeleton of *Cloudina* hints that a new race had begun: the life and death struggle between predator and prey. As many as a quarter of *Cloudina* fossils from certain localities have been found with holes in their skeletons, which could have been made by other organisms attacking and boring into them. Whatever the attacker was, it was a selective hunter. Similar shelled organisms living alongside *Cloudina* don't show the same damage. This is some of the first evidence of specialized predation in the fossil record, a force that has shaped the evolution of animals ever since.

Kimberella – First Bilaterian

***Kimberella* is an animal literally split in two, reflecting the most important major branching of animal life. This slug-like creature scratched its way across the rich microbial carpet of the Ediacaran seafloor. Although its family relationships remain unclear, thousands of exceptionally preserved fossils provide exquisite detail of its life, growth and death at the dawn of life.**

The oval-shaped body of *Kimberella* grew up to 15 centimetres (6 inches) long and looked like the bowl of a spoon, with a patterned outer rim and dappled upper surface. It lived on the shallow seafloor in what is now Australia and Russia, around 555 million years ago. Feeding on thick microbial mats in this calm and spartan environment, *Kimberella* thrived alongside other enigmatic Ediacaran organisms in Earth's increasingly fertile waters.

Among the many odd-looking creatures that made the seabed their home at this time, *Kimberella* is important for what it can tell us about the most basic divisions in the animal kingdom. As the first organisms experimented with their body plans, some of them gave rise to lineages that would go on to thrive and diversify in the Cambrian and beyond, while others never appeared again. With over a thousand fossils of *Kimberella* from multiple life stages, this organism is better known than many from the time period. It holds the key to tracing back these groups, uncovering the timing of life-changing moments in the story of life's evolution.

Mirror, Mirror

Kimberella was first thought to be a kind of jellyfish, but it is now suggested to be an ancient relative of molluscs. Evidence for this comes from strange scratch marks found near *Kimberella* fossils, which might be the tiny scrapings made by its mouthparts, called radula. Molluscs such as snails use their radula to remove and cut up their food, for example scraping algae from the surface of rocks. Although no radula is preserved in *Kimberella*, the scratches are a tantalizing hint that the ancient mollusc may have possessed one in life. The main part of *Kimberella*'s body is thought to have had a non-mineralized single 'shell', and the patterns around its rim could be the remains of muscle attachments for a single muscular foot. Taken together, this body plan would support the idea that it was a long-lost cousin of molluscs.

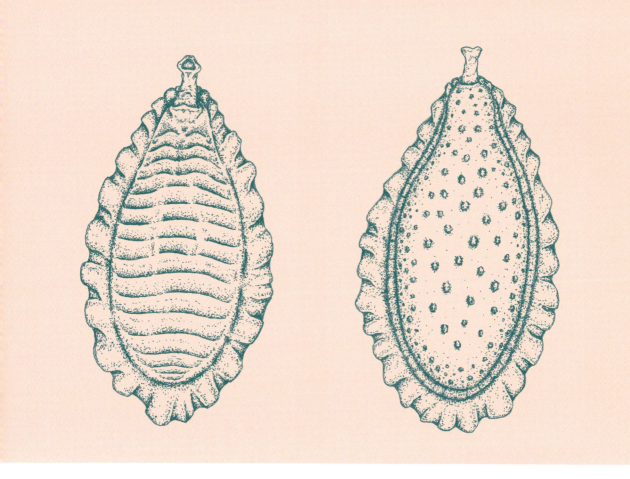

Whether *Kimberella* is related to molluscs is debated, but what most researchers do agree on is that it is the oldest known bilaterian. This is an animal with bilateral symmetry, where the right and left halves of the body are mirror images of each other. Bilaterians have digestive tracts with a separate mouth and anus. Most complex organisms on our planet are bilaterians, although some lose this symmetry as adults – for example, echinoderms such as starfish and sea urchins are bilaterally symmetrical as embryos but become radially symmetrical as adults. *Kimberella* marks a turning point in evolution, outlining the blueprint for most animal life on our planet.

PALAEOZOIC

The Palaeozoic era comprises six periods during which life made an unparalleled journey. From the first multicellular organisms on the seabed, animals colonized the land. Plants, invertebrates and finally vertebrates ventured from the water in pulses, creating new food webs and engineering revolutionary habitats. Wholescale changes in climate and the layout of the continents not only shaped the landscape but also drove evolution. Mountains and seas separated groups of organisms; landlocked heartlands were starved of water, while equatorial shorelines were lashed with tropical rain. As the world shifted in tantrums and doldrums, organisms adapted – or perished. This journey began 541 million years ago, at the end of the Ediacaran, and ended 289 million years later with destruction on such a massive scale that life was nearly wiped out entirely.

The evolutionary changes seen in life on Earth in the Palaeozoic were undoubtedly the most radical of all time. All of the animal groups and ways of life emerged for the first time in evolutionary history. Arthropods (invertebrates with an exoskeleton, segmented body and jointed limbs) grew massive in their marine kingdoms. Sea scorpions as big as motorboats hunted and scavenged in seas soupy with plankton. Vertebrate animals emerged, taking ownership of the oceans. On land, the first forests covered the landscape, initially dominated by giant relatives of club mosses and later by the gymnosperms – plants like conifers and cycads. Over time, the landscape swung from ice ages to humid swamp forests, to dry parched desert. By the Carboniferous, arthropods had escaped the water and evolved into giant insects as big as seagulls, their bodies fuelled by the highest concentration of atmospheric oxygen in Earth's history. In the Devonian, lobe-finned fish followed arthropods and took their first dry steps from shore. They soon separated into the three great lineages: amphibians, reptiles and mammals. The first amniotic eggs freed them from the water's edge, and by the Permian, they had exploded in number, size and lifestyle.

The Earth was populated by the predecessors of all the major animal groups we know today. By the end of the Palaeozoic, marine and terrestrial ecosystems had grown complex, filled with creatures we might recognize, as well as bizarre cousins that seem more science fiction than fact. It ended with a mass extinction worse than anything seen before or since. Complex life had taken a long time to emerge on our restless ocean world, but with the Ediacaran beta-test complete, in the Palaeozoic there was no stopping the innovations of nature from filling Earth with wonderful beasts.

Cambrian

541 to 485 million years ago. Abundant shallow seas provided perfect habitat for complex life.

The Panthalassic Ocean dominated the northern hemisphere.

The continents had a simple crust of micro-organisms, fungi and lichen.

Silurian

444 to 419 million years ago. The formation of the first soils allowed plants and then arthropods to colonize the land.

Northern continents were colliding to form Euramerica.

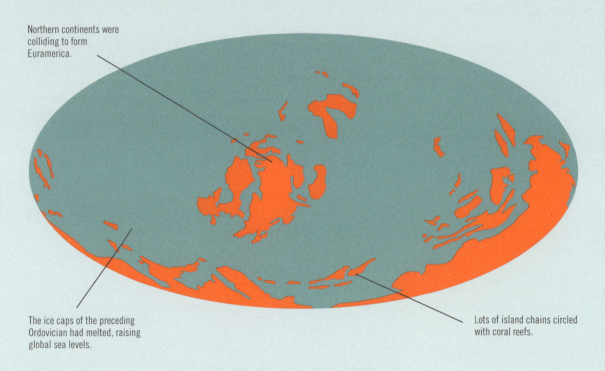

The ice caps of the preceding Ordovician had melted, raising global sea levels.

Lots of island chains circled with coral reefs.

Carboniferous

359 to 299 million years ago. There was 35 per cent atmospheric oxygen, the highest in life's history.

Most of the continents were covered in hot swamp forests, which formed coal deposits.

Glaciers formed at the poles for much of this time period.

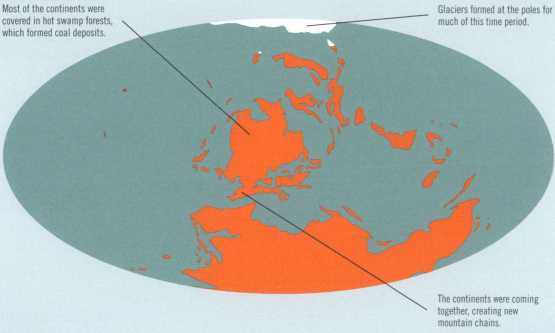

The continents were coming together, creating new mountain chains.

Permian

299 to 252 million years ago. At the end of the Permian, volcanic eruption in Siberia caused the largest mass extinction of all time.

The centre of the supercontinent was arid, while the shorelines were monsoonal.

The supercontinent of Pangaea formed.

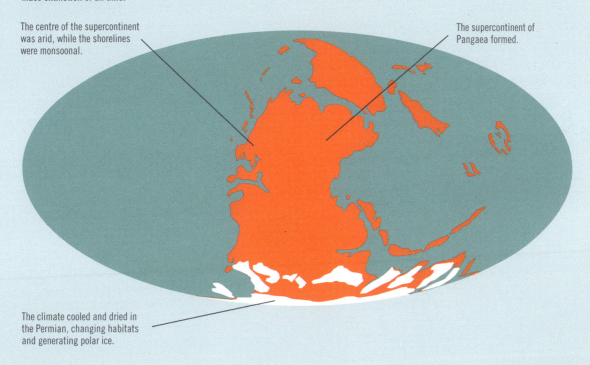

The climate cooled and dried in the Permian, changing habitats and generating polar ice.

Cambrian

A revolution on the seabed 541 million years ago in the Cambrian churned the ocean floor, kicking off the Palaeozoic. This heralded the origin of major animal groups, including the first definite invertebrates and vertebrates. Life, which had dallied long in the early days of the planet, suddenly morphed into countless forms, some even taking the first many-legged steps into the tidal margins of the land.

The first period of the Palaeozoic, the Cambrian, began 541 million years ago and lasted 56 million years. The Earth still looked much as it did before, but with the supercontinent Rodinia broken up, abundant new shallow seas formed, in which life thrived. What would become the northern continents separated from the rest, drifting across the vast Panthalassic Ocean. Gondwana – which comprised what are now Africa, Antarctica, Australia and South America – came together in the south. This massive continent cut off ocean circulation, preventing warm surface currents from reaching the South Pole and cooling global climate. It was still around 7 degrees Celsius (12.6 degrees Fahrenheit) warmer than the Earth is today, with little to no ice at the poles.

At the start of the Cambrian, a world-changing event took place on the seafloor. The microbial mats that had underlain earlier Ediacaran ecosystems like a dense carpet were ripped up by the first burrowing animals. This incited a revolution of ocean ecosystems, churning up nutrients and creating new niches to explore. Life erupted with newfound diversity, and thanks to exceptionally preserved fossils around the world, scientists know a great deal about these first animals. However, it was early days:

not much lived in the water column, with most animals still lurking on the seafloor. Arthropods such as trilobites are the most common fossils from the Cambrian, and were a major component of the ecosystem. There are some preserved with soft tissues as well as hard outer casings, and a few have been identified as the ancestors of insects and crustaceans. The predecessors of backboned animals began their evolutionary journey, growing the beginnings of a spinal cord and skull – the underpinnings of vertebrate life.

Plants had not yet evolved on land, and Earth's continents were still coated in a simple crust of micro-organisms, fungi and lichen. However, we know some Cambrian animals took their first ventures from the depths thanks to the footprints they left behind. Trackways like *Protoichnites* and *Climactichnites* tell us that at least some marine life was able to traverse the tidal flats. Rows of dots marching across the surface are probably the imprints made by arthropod feet, whereas slug-like molluscs drew lines in the sand.

Cambrian Explosion
It was once thought that all multicellular animals had their origin in the Cambrian. Their sudden appearance as fossils in

rocks from this time period was noted by European scientists in the 19th century, who eagerly declared it as evidence for the origin of life, referred to as the 'Cambrian Explosion'. We now know that there were complex animals in the preceding time period, but they finally become recognizable in the Cambrian.

There could be several reasons for the sudden emergence of so many new animals. Floating above it all, the ozone layer appeared for the first time in the Cambrian. This provided a shield of oxygen particles in the atmosphere, filtering out harmful radiation from the sun and protecting organisms from lethal damage. Movements in the Earth itself played their part: as continents shifted, volcanic activity and weathering of new mountain ranges could have affected ocean chemistry, increasing the amount of available calcium and phosphorous, and allowing animals to build mineralized skeletons. The evolution of sensory systems such as a centralized nervous system, body musculature and eyes may have driven diversification. The evolutionary arms race that began in the Ediacaran blasted off in the Cambrian. Predators honed their hunting skills, while their prey thickened their shells and found new ways to avoid becoming lunch.

Trilobites – Icon of the Palaeozoic

Trilobites are among the most recognized fossils in the world. These marine invertebrate animals lived in Earth's seas throughout the Palaeozoic. Thanks to their abundance and diversity, their fossils have been used for centuries to understand geological time and chart the processes of evolution.

Although now extinct, trilobites hung around on our planet for almost 300 million years, making them one of the most successful animals of all time. They are among the most common and recognizable fossils in the world, worn as amulets by indigenous people in North America and Australia, and among the first fossils to receive attention from European scientists. Trilobites were marine creatures that lived in the sea. They had a hard exoskeleton comprising three sections (hence their name, which means 'three lobes'), and it is this that survives in the fossil record. Trilobites were incredibly versatile, including filter-feeders, predators, those that remained on the seabed and those that could swim through the water column. They dwelled in both shallow and deep water all over the world, and grew to a huge range of sizes: the smallest being a few millimetres in length, while the largest were over half a metre (17 inches) long and would have weighed as much as a well-fed cat.

Having emerged in the Early Cambrian, trilobites are one of the first groups of arthropods – invertebrates with a segmented exoskeleton and jointed limbs – to evolve on Earth. Their exoskeleton moulted regularly as the animal grew. Sometimes only these discarded parts are found in the rock record, left where they were shed, like dirty laundry. Trilobites varied enormously in shape, sporting spikes and horns for defence or fighting. Their eyes were often complex, and had hard lenses, but these animals also had softer body parts, such as gills and antennae, which are very rarely preserved, and their internal organs are virtually unknown. Trilobite fossils are so abundant throughout the rocks of the Palaeozoic that they have even revealed mechanisms of geology such as continental drift, and the deep time processes of evolution.

Trilobites appear quite suddenly in the Cambrian fossil record, as though dropped from space. This suggests their initial evolution took place rapidly. They may have had ancestors in the Ediacaran, but this is uncertain. Their tough exoskeleton preserves well, which is why they are so common in Palaeozoic rocks. There were over

The trilobite, *Homalonotus armatus*, is an extinct arthropod.

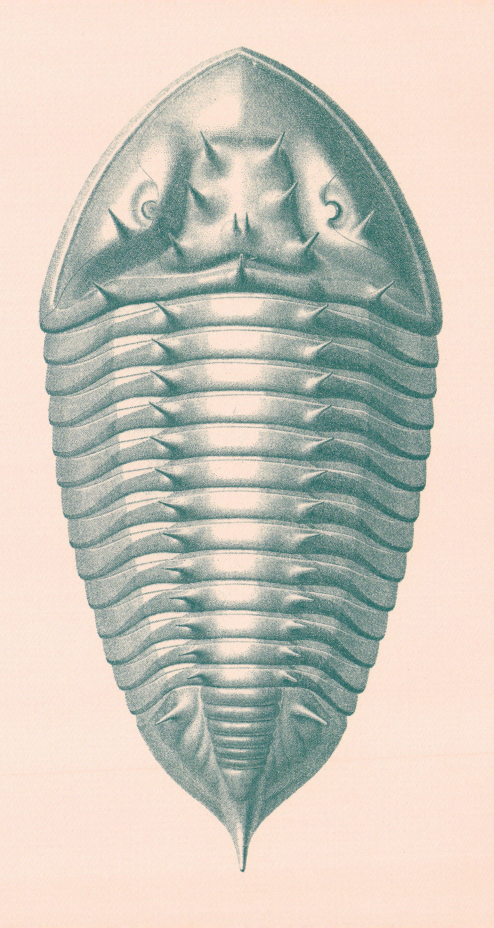

20,000 species found all over the world, and their numbers and diversity mean they can be used to define different time periods, marked by the presence or absence of different species. Fossils that can be used in this way are known as index fossils. Most of the rock layers of the Cambrian are marked by changes in trilobites, like a fossil clock ticking through deep time.

End of an Era

Although trilobites are iconic symbols of the science of palaeontology, this group of animals lived only in the Palaeozoic. They thrived in the Cambrian and Ordovician – the first two periods of the Palaeozoic – but by the Devonian, they began to decline. They finally died out in the biggest mass extinction of all time, at the end of the Permian, before the appearance of the dinosaurs.

Thanks to their beautiful shape and presence in our ancient rocks, humans have treasured trilobites for millennia. A trilobite fossil was found by archaeologists in the caves of Arcy-sur-Cure, in France. It appears to have been worn as a pendant over 15,000 years ago, and has been rubbed smooth from handling to the point that the species can no longer be identified. References to trilobite fossils are found in ancient Chinese manuscripts, where they were prized for their form and beauty. The Greeks and Romans also discuss their uses, and some indigenous people in North America kept them as amulets or sacred objects; the Hopi and Crow often placed them in their medicine bundles. Trilobite fossils remain a staple of the fossil trade, sold in gift shops around the world and incorporated into jewellery to this day. Trilobite are used in the logo for countless brands and are an enduring symbol of palaeontology, and the beauty and antiquity of our living planet.

Trilobites came in many shapes and sizes, making them ideal as markers for changing time periods.

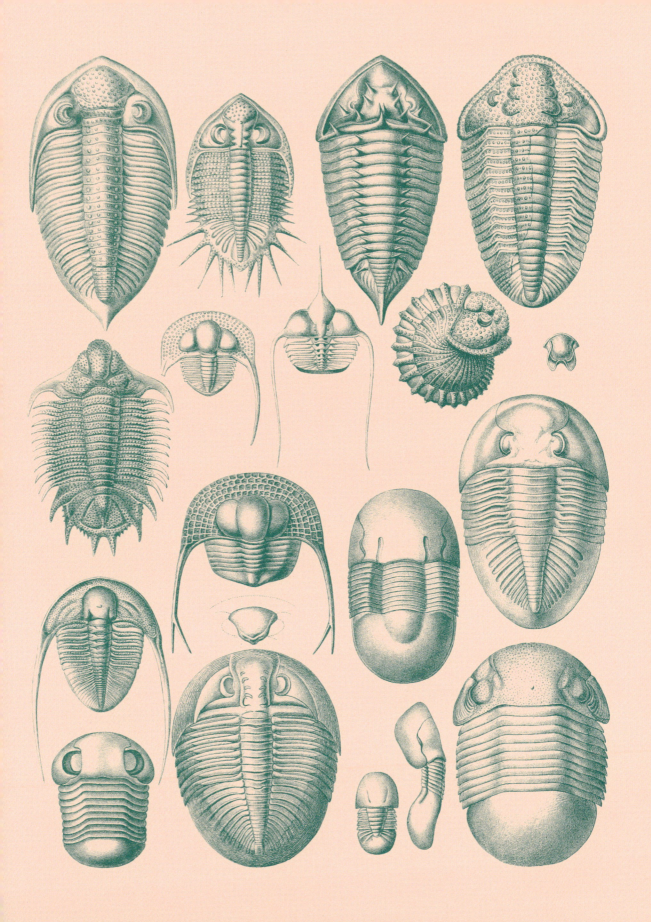

Myllokunmingia — First Vertebrates

The earliest relatives of all backboned animals swam in burgeoning Cambrian seas. They include creatures like *Myllokunmingia*, a small, spear-shaped creature discovered in rocks from China. It was among the first animals to possess the beginnings of a spinal cord and other characteristics that provide the blueprint for vertebrate life.

Myllokunmingia resembled a floating leaf. At just 2.5 centimetres (1 inch) long, it was a tiny creature. It lacked jaws and probably filter-fed on passing plankton. Although humble, this creature possessed the underlying architecture for all backboned animals alive today: the precursor of the spinal cord, called a notochord; a well-defined head and tail; paired sensory structures, including eyes; a dorsal fin; and a body made of repeating muscle blocks. The fossils of *Myllokunmingia* have been found in Yunnan Province in China. They are preserved in extraordinary detail, revealing that it had six gills and distinct zigzag muscle segments along the body. Although we don't know precisely where *Myllokunmingia* sits in the animal family tree, it is certainly the oldest vertebrate animal known, hinting at how our body plan was assembled at the beginnings of life on Earth.

Older Than We Thought

The discovery of *Myllokunmingia* pushed back the timing of the evolution of vertebrate animals. Researchers thought animals with backbones were relative latecomers compared to the invertebrates. It now appears backboned animals were also part of that sudden rocketing of life that took place in the early part of the Cambrian. This tells us that the forces that drove the Cambrian Explosion were universal, and happened exceptionally quickly.

Animals like *Myllokunmingia* can only whisper their ancient secrets thanks to their unusually good preservation. Usually, only mineralized parts such as shells and bones are fossilized, but there are circumstances that allow even the delicate soft tissues such as skin, hair and internal organs to survive deep time. Rocks that hold such fossils are known as Lagerstätten, from the German for 'storage place'. The conditions that create a Lagerstätte vary, but usually organisms are buried by fine sediments in an oxygen-poor environment (anoxic), slowing their decomposition. There are over 75 recognized Lagerstätten, and at least 11 are Cambrian. Thanks to these, we have a great deal of insight into the early evolution of animal life.

Myllokunmingia was a chordate that lived around 520 million years ago, and was one of the earliest vertebrate animals.

44

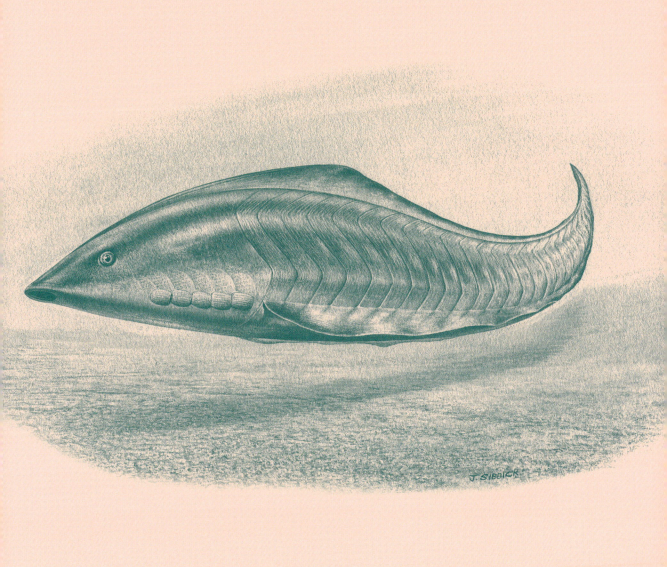

Ordovician

Planet Earth in the Ordovician was predominantly a water-world. New types of animals continued to evolve in the shallow seas, many sheltering in the first coral reefs. On land, plants festooned riverbanks and seashores, turning the continents green. The period ended with the first of the six biggest mass extinctions in Earth's history, redirecting the course of evolution.

The Ordovician is the second time period in the Palaeozoic, beginning 485 million years ago and lasting for 41 million years. The pace of evolution had slowed after the Cambrian Explosion, but in the Ordovician it was reinvigorated in the grandly, if not imaginatively, named 'Great Ordovician Biodiversification Event'. This signalled the end for many earlier groups of organisms, replaced with a rich new fauna that sailed through the water column of our endless oceans. Sea levels rose, reaching the highest levels in the whole Palaeozoic. The Panthalassic Ocean still dominated the planet, but smaller seas such as the Proto-Tethys and Iapetus sloshed between the drifting continents, creating shallow marine habitats. In the last part of the Ordovician, sea levels fell again and temperatures cooled – leading to a cataclysmic ice age and mass extinction.

Evolution continued to drive revolutionary adaptations in organisms. In the plant world, the first vascular plants (those with internal channels for water and nutrients to flow through) developed, while in the sea, coral colonies were restructuring habitats on the seabed. Invertebrates such as arthropods continued to proliferate, and included more shelled animals such as brachiopods and molluscs. Among the vertebrates, the first jawed fish evolved – a vital development in their evolutionary history.

It is thought that there were a hundred times more asteroid impacts in the Ordovician than in more recent geological periods. These varied from small scattered fragments to collisions comparable to hydrogen bombs, and they had far-reaching effects on the local and global environment – scientists even speculate that the cosmic knocks played a role in driving natural selection.

Great Biodiversification Event

The diversification of animals in the Ordovician was a pivotal moment in the history of life. The number of major animal groups tripled, and the new ecosystem they created was rich in filter-feeders. Plankton became increasingly species-rich and widespread; this brew of tiny organisms still underpins ocean food chains around the world. Animals called graptolites, previously restricted to feeding on the seabed, took to the open sea to net their miniscule plankton prey. Whereas before, animals had tended to be similar across the globe, the patterns of life became more localized as creatures evolved new ways to feed and defend against predators. The first coral reefs were constructed, and these underwater cities created novel living space for other organisms to occupy.

Mass Extinction

Earth's evolutionary history is punctuated by multiple mass extinctions. Six of these have been identified as the most significant in terms of numbers of species lost and global scope. The first of these brutal extinctions occurred at the end of the Ordovician, caused by an ice age. The ice sheets were centred on what is now the Sahara Desert, which lay near the South Pole at that time. The increase in glacial ice lowered sea levels, leaving swathes of ocean habitat high and dry, and cooled the oceans to 5 degrees Celsius (9 degrees Fahrenheit) below the temperature they are today. Global temperatures were some of the lowest since Snowball Earth. This ice age could have been triggered by increased dust in the atmosphere from asteroid impacts, which would have blocked the sun's warmth. New types of plants and the weathering of rocks also removed the greenhouse gas carbon dioxide from the atmosphere, adding to the cooling effect. Around 61 per cent of marine life became extinct. This process occurred more than once, with each deep-freeze cycle causing further extinctions. Despite all of this, life found a way through.

Coral — The First Reefs

Corals are a crucial component of the Earth's marine ecosystems. The first coral reefs grew in the Ordovician, constructing a new habitat for ocean wildlife. Their growth drew carbon from the oceans and atmosphere, changing Earth's geochemistry. Their fossils are found around the world, comprising unfamiliar shapes belonging to species unknown today, but with an equally important role in their ancient ocean world.

Corals often look more like rocks than living organisms. Hard corals are made up of sack-like animals that secrete a hard mineral exoskeleton, while soft corals lack this solid skeleton. Many corals live in a symbiotic relationship with photosynthetic organisms, which produce nutrients in exchange for safety and feed upon the coral's waste. Other corals filter-feed on plankton, or catch small fish. They are ecosystem builders, intimately tied to the organisms that live around them.

Surprisingly, corals belong to the same group of animals as jellyfish and sea anemones, collectively called cnidarians (nai-*dare*-eeans). They first appeared in the Cambrian, but it was not until the Ordovician that they became abundant. The first colonies were built by species unknown today: the rugose and tabulate corals. Both died out by the Early Triassic, replaced by the stony and the soft corals we have today. Of the copious sedimentary rocks on our planet, corals form deep layers chock-a-block with the remnants of their thriving world. Not only are their mineral exoskeletons preserved, but also the remains of other marine life that formed part of the reef, such as sponges, echinoderms and shelled animals such as molluscs and brachiopods.

The First Reef Builders

The first corals created their exoskeletons from calcite, a mineral that fossilizes easily. This makes them excellent index fossils, used for interpreting the age of rocks. Modern coral groups are composed of aragonite instead, which isn't as readily fossilized. As a result, we know far less about their evolution, even though it occurred much more recently.

Rugose corals were shaped like horns, with intricate exoskeletons made of repeating plates. They lived at different water depths, some solitary, others forming large colonies. The colonial forms were often relatively small, with each one perhaps only a few centimetres

Rugose corals *Columnaria alveolata* (top) and *Lithostrotion basaltiforme* (bottom).

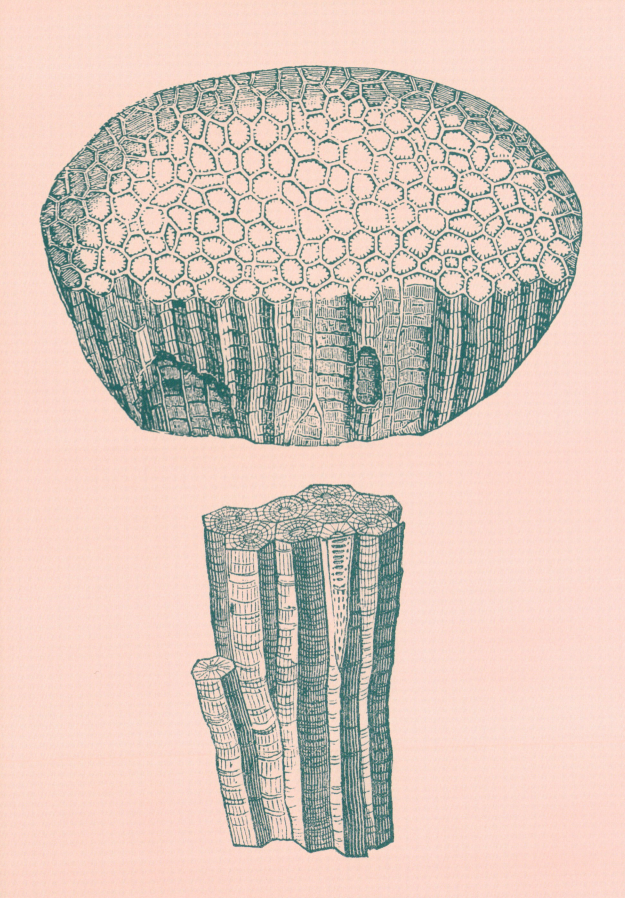

in length, whereas solitary species could reach almost a metre (3 feet). Tabulate corals were smaller than rugose corals, and formed colonies with intricate honeycomb-shaped hexagonal exoskeletons. Their reefs could be flat, spherical or cone-shaped, and they tended to form in shallower water.

Reefs in Danger

Coral reefs are a vital part of the Earth's planetary systems. Their growth alters the geochemistry of the oceans and atmosphere by removing carbon, and they provide habitats for marine animals, particularly as safe nurseries for millions of fish and invertebrate species. Coral provides important economic benefits too, replenishing global fishing stocks, as well as through tourism, as people travel the globe to immerse themselves in the beauty of a flourishing reef.

They are also fragile. Over half of today's coral reefs face extreme threats from climate change, habitat destruction and pollution. Coral bleaching – when the photosynthesizing organisms that dwell in corals die or are expunged, often due to extreme increases in temperature – decimates reefs. Although coral can survive short periods without their symbiotic partners, they receive up to 90 per cent of their energy from them, and so bleaching is usually fatal. It is estimated that 10 per cent of global coral reef ecosystems have already been lost, and that as much as 50 per cent may be gone over the next decade if we don't curb our destructive impact on the natural world.

Rugose coral, *Goniophyllum*
(top), and tabulate coral
Halysites catenularis (middle
and bottom).

Graptolites — Open Ocean Pioneers

Graptolite fossils look like cryptic pencil scratchings, but they are all that remain of floating colonies of thousands of individual animals. In the Ordovician they were one of the first animals to exploit the open ocean, drifting through the seas feeding on plankton. Their traces provide high-precision dates for rocks, and have helped scientists reconstruct the complex geology of the past.

You could be forgiven for overlooking the fossils of graptolites when searching rocks from the Palaeozoic. Many don't look like fossils, appearing to have been drawn on to the surface of rocks. This led Carl Linnaeus to comment that they were 'pictures resembling fossils', rather than the real thing, naming the first graptolite *Graptolithus*, which means 'written rocks'. Some resemble open zips, others look like leaves or fragments of feather. These shapes are made by the collagenous framework of the graptolite, which would have hosted up to five thousand individual animals. Although these tiny residents, called zooids, are not commonly preserved, it is thought that they filter-fed from the surrounding sea using minute comb-like snares to snag passing plankton.

Graptolites first evolved around 520 million years ago in the Cambrian, and died out in the Carboniferous, around 180 million years later. But it was in the Ordovician that they had their epic heyday. From unassuming sedentary animals on the seabed, they grew to become the world's first open-ocean sailors. Most graptolite fossils are found in shales or mudrocks, formed at the bottom of the deep ocean, and they are so numerous, with clear changes in their shape and structure through time, that they provide biomarkers for geologists to date rocks accurately. As their shapes also varied with environmental conditions, they can tell us the water depth and temperature at particular locations, making them incredibly useful for understanding the ocean geography of the past. Graptolites are thought to have been hermaphroditic – possessing both female and male reproductive organs. They may have alternated their sex, or changed sex with age or their position in the colony, no one knows for sure. Although their position in the tree of life remains uncertain, most researchers consider them to be relatives of echinoderms such as the sea urchin.

A selection of graptolite fossils, demonstrating their diverse shapes.

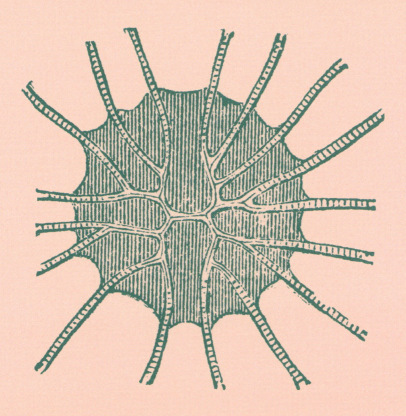

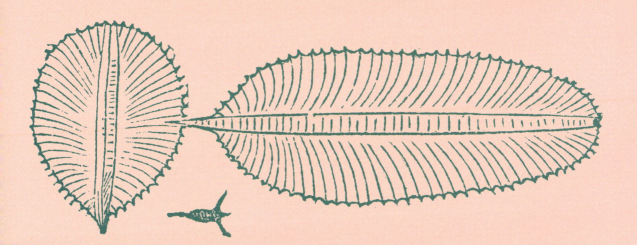

Roaming Hordes

The first graptolites were quite unlike the free-floating colonies that would later drift through the ocean. In the Cambrian they were rooted to the seabed and resembled veined leaves. Some bottom-dwelling graptolites lived by encrusting other organisms and rocks, or were rooted to the seafloor. But at the end of the Cambrian, everything changed: in the aftermath of an extinction event, the marine ecosystem was restructured and graptolites were among the animals that responded radically. Their free-swimming larvae, rather than sinking to the seafloor to complete their life cycle, joined forces in the water column like formation skydivers, constructing colonies. How these colonies manoeuvred is unclear, they may have been carried along on currents, but it is possible that the graptolites steered their course and moved up and down the water column using undulating motions of their appendages.

In the Silurian Period graptolites evolved into increasingly intricate structures. These could be coiled like an extravagant eyelash, form tiny dizzying spirals, or be a long straight series of lobes. It is the diversity of their shapes that makes them so useful, and inspires fierce scientific passion of those who study them. Free-floating graptolites finally died out in the early Devonian, and although seafloor-dwelling species persisted a little longer, by the Carboniferous, they too became extinct.

Amazingly, some graptolite fossils are found in metamorphic rocks that have been squeezed and distorted by tectonic forces. When such rocks are folded and stretched, the fossils are deformed with them, resulting in almost unrecognizable imprints. These shiny streaks threading through the rocks bear witness to the colossal forces that continually reshape our planet.

Dendroid graptolite,
Dictyonema cijassibasale,
which lived on the seafloor.

Conodonts – First Vertebrate Predators

Throughout the planet's Palaeozoic marine rocks are scattered the tiny fossilized mouthparts of conodonts. Misunderstood for over a hundred years, these important index fossils belonged to one of the first vertebrate predators on Earth: an eel-like creature that existed in our seas for over 285 million years.

For over a hundred years after they were first discovered, conodonts were only known as tiny little mineral shapes. Like microscopic snow-flakes, they came in a plethora of stunning forms: hooks and fans, combs, seed-like blobs, stars and knobbled whorls. Smaller than grains of rice, these strange fossils appeared in the Cambrian and speckled Earth's rocks all the way through to the Triassic, leaving their cryptic traces behind like breadcrumbs.

Only when an exquisite fossil conodont was found with soft tissues preserved did palaeontologists finally discover what all of these odd shapes actually were. They turned out to be the mineralized mouthparts of an animal that resembled a jawless eel. Some species were no longer than a fingernail, others nearly half a metre (20 inches) in length. What's more, they were vertebrates, adding to our picture of the evolutionary emergence of backboned animals. They died out only when sea-level changes, ocean acidification and the rise of other new marine organisms in the Triassic proved too much for them. Thanks to their longevity as a group, these incredible creatures are important index fossils for interpreting deep time and the history of rock formations over millennia.

Microscopic Mysteries

The first conodonts were identified in 1856 by the Baltic German biologist Heinz Christian Pander (1794–1865). They were known only from their tiny hard parts, composed of similar minerals to our teeth and bones. Many conodont microfossils are so miniscule – as little as 200 microns (0.2 millimetres) – you could fit a dozen of them on the head of a pin. In the years that followed their discovery, scientists tried to figure out what kind of creature these little structures belonged to. Most recognized that they were either mouthparts or claws, perhaps belonging to extinct worms or snails. Others suggested they might even be from a kind of plant.

It wasn't until over 120 years later that a fossil animal was found in Scotland with soft tissues preserved, and the mineralized

Conodonts are now known to be eel-like animals, and include the first vertebrate predators in the world.

mouthparts in place. More of their bodies were soon revealed in Lagerstätten (rocks with exceptional fossil preservation) from the United States, South Africa and beyond. These new fossils showed not only that it was an elongate animal, but also that it was a vertebrate, because it possessed a notochord, zigzag muscle arrangements along the body and an asymmetrical tail fin. They also had a pair of eyes that could rotate using strap-like muscles, a feature only seen in vertebrates.

The incredible diversity of conodont mouthparts suggests that they were experimenting with their diet. Some were probably fil-ter-feeders, swimming through the plankton with mouths agape. Others actively hunted their meals. The teeth-like structures of conodonts bear microscopic wear patterns that show they were grasping, shearing and grinding their food. This makes conodonts some of the first vertebrate predators in the world – but certainly far from the last.

A selection of conodont mouthparts, which litter the early fossil record like snowflakes.

Silurian

The Silurian is the shortest but most frenetic period in the Palaeozoic. The icecaps melted and raised sea levels, while continents collided to erect the oldest mountain chains on Earth. The first vascular plants evolved, and arthropods set their copious feet on land. Meanwhile in the seas, fish developed mineralized skeletons and jaws, opening novel ways of life for them and all of their descendants.

The Silurian lasted just 25 million years, ending 419 million years ago – a heartbeat in geological time. It began in the aftermath of the Ordovician–Silurian mass extinction, which killed over half of all marine life. Although it is the shortest time period in the Palaeozoic, this intense moment in Earth's history captures some of the most important events in the evolution of our planet.

The icecaps left over from the end-Ordovician ice age shrank in the Silurian. As the water flowed back into the ocean, it raised sea levels and inundated the fringes of the continents, creating island chains encircled by bustling, colourful reefs. The great continent of Gondwana still dominated the southern hemisphere, while the scattered fragments of the northern continents coalesced to form a continent called Euramerica. As the plates collided they crumpled and deformed, thrusting up immense mountain chains hundreds of miles long. This is known as the Caledonian Orogeny, and the remains of this globe-spanning range still exist today along the east coast of North America (the Appalachians), Ireland, Scotland and Scandinavia. Although rubbed as smooth as cue tips by over 420 million years of erosion, they still

dominate the landscape, and their presence has influenced human settlement and culture.

Towards the end of the Silurian, the climate warmed further – a trend that continued into the Devonian. As the surface of the ocean heated and sea levels fluctuated, violent storms whipped up and battered the coastlines of our planet, creating challenging conditions for life. The evidence for this destructive weather is seen in the rock record: beds of broken shells and shattered reefs, left in the wake of raging tropical typhoons.

Legs on the Land

In the Silurian, invertebrates permanently set their multiple tiny feet on dry land. These pioneers were only a few centimetres in length, and included the ancestors of myriapods – centipedes and millipedes – as well as relatives of spiders and scorpions. Their move was only possible thanks to the plant life that had already made a home there. Lichens, algae and fungi had colonized the near-barren Earth by the Silurian, along with the first plants with circulatory systems for water and nutrients. Although scattered fossil spores found in Ordovician rocks hint that such plants may have appeared earlier, it is not until the Silurian that we find definitive fossils. These alien-looking structures are preserved in exquisite detail, providing a vivid glimpse of the ancient surface of our planet. This was a landscape being rapidly reshaped by the animals and plants that were colonizing it and building new ecosystems.

Intricate Food Webs

After being hit hard by the mass extinction at the end of the Ordovician, marine life flourished again in the Silurian. Food webs became increasingly complex, weaving intricate links as groups proliferated into new shapes and sizes. The many limestone rock formations from this period reveal the bountiful growth of corals, sponges and bryozoans, which fed on passing swarms of plankton and small invertebrates in the water column. Apex predators called eurypterids, which looked like a cross between a scorpion and a lobster, grew to immense sizes, becoming the largest arthropods to have ever existed.

Vertebrate animals were also going through foundational changes in their evolutionary history in the Silurian. The first fish made their way up rivers and inhabited lakes and streams, giving rise to freshwater lineages. Fish also evolved the first mineralized bones and jaws – with life-altering consequences. Possessing jaws to bite and crush allowed them to occupy new niches in their ecosystems, and consequently drove evolutionary responses in their prey. Among the shoals living at this time were the fishy ancestors of all backboned animals today, ourselves included.

Prototaxites – Terraforming Soils

In the Silurian there was a life form on our planet unlike anything before or since. *Prototaxites* was a giant fungus as tall as a house. By far the largest organism alive at the time, it helped shape the first soils, a prerequisite for life on land. *Prototaxites* persisted for over 100 million years before disappearing forever, leaving only its smaller fungal relatives to flourish to the present day.

When we think of fungus, we imagine small mushrooms pushing up from a forest floor. But 425 million years ago, there was a towering fungus that populated our planet before the first animals emerged onto land. *Prototaxites* was shaped like a branchless tree trunk, projecting over 8 metres (26 feet) into the air. Taller than two giraffes and around a metre (3 feet) in width, it was by far the largest organism on Earth for millions of years.

We have fungi to thank for all life on the Earth's surface. They are one of our pioneering terraformers, creating soil for other plants and animals to live upon. One of the oldest fossil fungi on land is *Tortotubus*, which appears in the Ordovician and resembles very fine hairs. Fossil fungus is hard to tell apart from microbes due to its small size and lack of a skeleton, but *Prototaxites* is an obvious exception, dominating the ancient landscape. When it was first discovered in Canada in the 19th century, it was mistakenly thought to be a tree trunk due to the concentric rings of growth – the name *Prototaxites* means 'first yew'. At different times over the last 100 years, *Prototaxites* was thought to be an algae, plant or fungus. It wasn't until the fossils were re-examined in the 1990s that most researchers agreed it was a type of fungus, perhaps even a lichen.

The first fungi not only helped create soil, but provided food and shelter for the first life on land. Tiny bore-holes in some *Prototaxites* fossils are suggested by some researchers to be the result of insects chewing their way into the organism to feast and nest. The growth rings in the 'trunk' are lopsided and filled with tube-shaped cells, which resemble fungal hyphae. It's not clear why *Prototaxites* became extinct, but it could have been competition for space and nutrients with newly evolving plant species. By the end of the Devonian, this giant had disappeared, but its relatives continue to thrive everywhere on Earth, especially in damp places.

A reconstruction of *Prototaxites loganii*, with a wrinkled outer surface and branching structures.

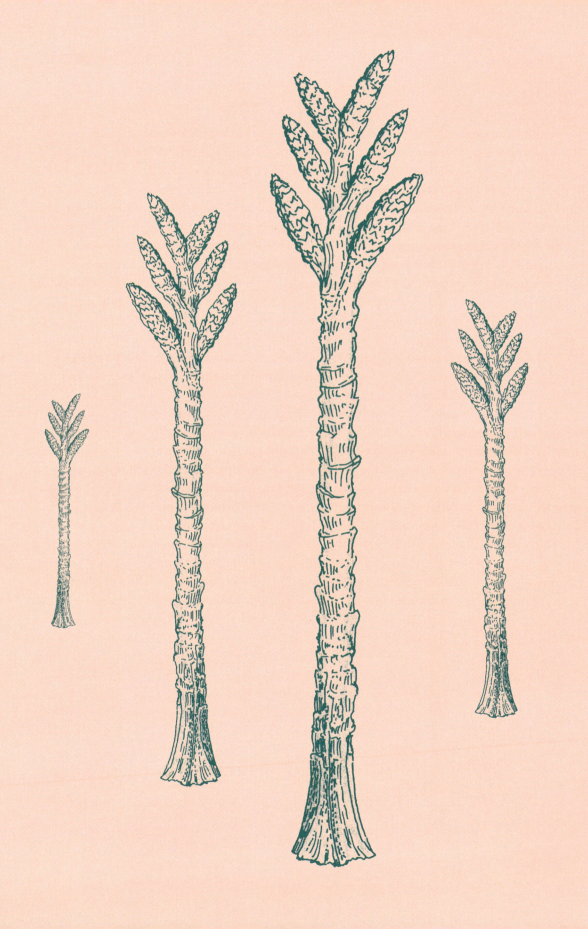

Feasting on the Dead

Fungi might look like plants, but they belong to their own separate branch of the tree of life. What's more, molecular analysis tells us they are more closely related to animals like ourselves than they are to plants. Fungi are heterotrophic, meaning they must obtain nutrients from other organisms. They often feed on dead and decaying organic material, but others are pathogenic, causing disease in their host (such as ringworm or Valley Fever).

The first fungi on land pose an enigma, because it's not clear what they fed upon. For this reason, lichen is thought to be one of the earliest land colonizers, thanks to its partnership with photosynthesizing algae and cyanobacteria. However, *Prototaxites* may have been a decomposer, feasting on the earliest organisms on the planet. Studies of isotope ratios in its tissues suggest *Prototaxites* obtained carbon from other living tissues, rather than by photosynthesis. This evidence is controversial, and the jury is still out about how this strange giant lived and fed.

Fungi undoubtedly played a crucial role in the transformation of the Earth's surface. Fungal growth, combined with erosion by water, wind and ice, would have helped create the first soils for plants to colonize. They have the ability to break down almost anything using digestive enzymes, and are vital for recycling nutrients and material back into food webs. It has been suggested that there may be an increase in fungi after mass extinctions, when they are hard at work breaking down the debris of extinct life.

Fungi to be Around

Algae and fungi probably colonized the land together, with fungi feeding on the algae, or in symbiosis with it, in relationships that may go back over a billion years into the Precambrian. A fossil called *Ourasphira* found in the Canadian Arctic is thought to be among the earliest evidence of fungi on our planet.

Fungi have formed relationships with many different plant and animal groups in their long evolutionary history. Ants and termites famously 'farm' fungi inside their nests, with fungi and insect unable to survive without each other. One of the most important relationships between fungi and other organisms is that of mycorrhizal symbiosis with plants. Mycorrhizal fungi are visible in soil as fine white filaments called hyphae, around the roots of a plant. They release phosphorous and other minerals crucial for plant nutrition, and their hyphae stabilize soil and retain moisture. This close tie between fungi and plants is thought to have existed for over 400 million years, and today over 85 per cent of all plant species foster this special relationship. Without fungi, it is unlikely that plants, and the complex terrestrial ecosystems that they underpin, would ever have evolved on our planet.

Cooksonia – First Vascular Plants

Cooksonia **is one of the first plants in the fossil record, and the first vascular plant, with specialized tissues for transporting water and nutrients. By utilizing energy from the sun and water from the soil, vascular plants turned the world green and shaped new ecosystems. They were predecessors of most plant groups alive today, transforming Earth's terrestrial landscapes forever.**

The evolution of plants is one of the most important stories in Earth's history. They play a vital role in configuring our atmosphere through gas exchange, alter geochemistry through their growth and decomposition, underpin food webs and create habitats for other organisms. Plant fossils are rare because they lack mineralized tissues. This makes it difficult to trace their emergence and evolution through time. The earliest plants are often represented by their spores alone – such fossils in the Ordovician suggest plants had begun to colonize land by that time. But the earliest 'body' fossil of a plant belongs to *Cooksonia*. It grew in the late Silurian and remained an important part of the planet's flora for many millions of years.

Cooksonia stood just a few centimetres tall and had thin stems topped with trumpet-shaped reproductive structures, packed with tiny spores called sporangia. They grew from a creeping stem called a rhizome, rather than roots, and had no leaves or flowers. Although likely green in colour, *Cooksonia* probably didn't rely solely on photosynthesis for nutrients. It is important because it is the earliest known vascular plant, the group that dominates our planet to this day.

Ecological Succession

Living communities change over time in a process called ecological succession. This can happen on scales from hundreds to millions of years. Communities form around a few pioneering organisms and become increasingly complex over time. Often they reach a stable point, called the climax community, and remain in balance until a disturbance alters conditions, for example natural disasters such as wildfires or landslides. Looking at the fossil record, we can trace ecological succession across deep time as new land is colonized, groups evolve and extinctions occur on small to global scales, reshaping the world.

The landscape of our planet was colonized first by pioneering bacteria, then algae and fungi. Land plants evolved from aquatic

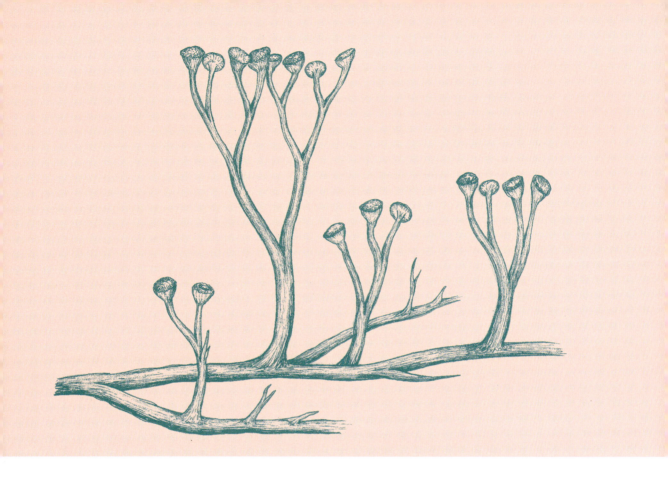

Cooksonia was among the first vascular plants to evolve on Earth.

green algae, and would have initially festooned the edges of lakes and rivers, close to water. There, it helped generate the first soils, which were crucial for creating the conditions for new kinds of plants to evolve. As they developed a vascular system and roots, these plants became less reliant on surface water and spread to new habitats and drier regions, accessing water from deeper in the soil, and moving it throughout their tissues. Eventually, where the conditions were right, the bare rocks and dust of our planet turned green.

Eurypterus – Biggest Arthropods on Earth

Eurypterus lends its name to one of the most prolific groups of hunters in the Palaeozoic: the eurypterids, or sea scorpions. They included the largest arthropods to ever live on Earth, prowling aquatic environments, from the dark seabed to shallow inland swamps. Equipped with powerful paddles and pincers, *Eurypterus* and kin were formidable predators for an astonishing 200 million years, before vanishing forever.

Eurypterus lived over 420 million years ago in the oceans of the northern hemisphere. It was a type of sea scorpion, or eurypterid, which resembled a lobster with a hard exoskeleton and large paddles at the front of the body, probably used for propelling it through the water like a pair of submarine oars. This group appeared by the start of the Ordovician, but it was in the Silurian that they became a major part of ocean ecosystems. They were swift marine predators that could tear apart their prey with powerful pincers.

Sea scorpions are arthropods, the group that includes invertebrates with a segmented body and jointed appendages, such as insects, spiders, millipedes and crustaceans. Although they resemble scorpions or lobsters, eurypterids are thought instead to be more closely related to horseshoe crabs (Xiphosura) and spiders (Arachnida). They were an astonishingly prolific and long-lived group found all over the world, in marine, brackish and fresh water. Some researchers have even suggested that they may have been able to come onto land, thanks to a dual respiratory system that processed oxygen from air as well as water.

Eurypterids had compound eyes oriented to the front of their body, giving them stereoscopic vision for targeting prey. To other life forms dwelling in our planet's ancient seas, these sea scorpions would have been an ever-present danger. *Jaekelopterus,* for example, was a giant eurypterid that lived in the Early Devonian, and grew to 2.6 metres (8.5 feet) from head to tail – longer than a super-king-sized bed – making it the largest arthropod in Earth's history. Although there were several other giant species, most eurypterids were much smaller, often less than the length of your hand, and the smallest was no bigger than a grape. Despite flourishing for over 200 million years in our seas, sea scorpion diversity declined after the Silurian, and the group disappeared in a devastating extinction event at the end of the Permian.

Eurypterus remipes, a eurypterid or sea scorpion from the Silurian Period.

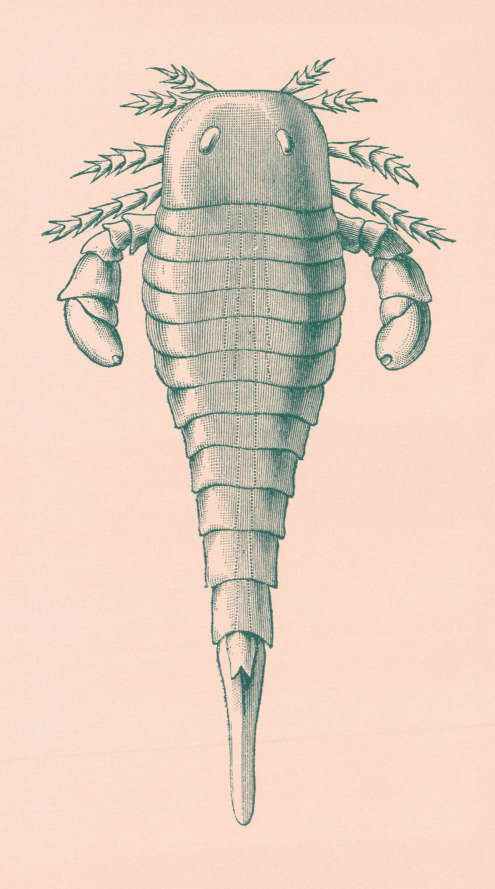

Claw Horn Clan

Eurypterids are part of a group of arthropods called Chelicerata, meaning 'claw horn'. They include horseshoe crabs, arachnids (harvestmen, mites, scorpions and spiders) and sea spiders. Chelicerates have a history stretching from the Cambrian to the present day. Their name comes from their first appendages (often near, or part of, the mouth), called chelicerae, which look like fangs or pincers. They are one of the major groups of arthropods on the planet, playing vital roles as predators and scavengers in our ecosystems.

Although the group now includes many land dwellers – and their fossils include some of the first animals on land – chelicerates first evolved in the sea. Horseshoe crabs (*Limulus*) are one of the closest relatives to eurypterids. These crabs are famously nicknamed 'living fossils' because they appear unchanged since the Silurian. However, thanks to the process of natural selection, no animal remains static, even if superficially their outer appearance changes very little. Modern *Limulus* are not the same species as their ancient horseshoe crab cousins, and their body plans differ in significant ways.

Cost of Gigantism

Giant sea scorpions such as the gargantuan *Jaekelopterus* faced a physical cost for their impressive size. Like all arthropods, eurypterids had a tough cuticle covering their bodies, which couldn't stretch as they grew. They therefore moulted regularly, wriggling out of their old exoskeleton and discarding it like old clothing. This meant that giant eurypterids lost a lot of invested energy during moulting. They also struggled to absorb enough oxygen to fuel their hefty body mass, and their bulk meant they moved more slowly.

To compensate for all of these drawbacks, large eurypterids tended to have extremely thin, unmineralized exoskeletons. As a result, the main parts of their body are less commonly fossilized, because they were more likely to decompose or be otherwise destroyed by the processes of deep time. The few delicate outer coatings that are preserved are paper-thin, which would have made their bodies lightweight in comparison to their size, reducing the cost of moulting. Although their bodies were delicate, their spiked claws certainly weren't – they remained sturdy, enabling even the most scantily clad giant to effortlessly dismember its prey.

Eurypterus remipes, a eurypterid from the Silurian Period, pictured from underneath.

70

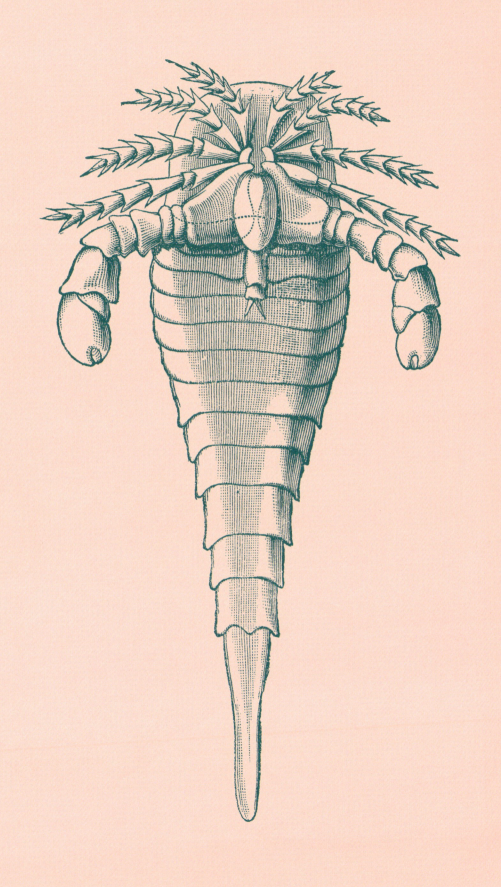

Devonian

The Devonian spans 419 to 359 million years ago, and is often called the 'Age of Fish'. Some of the most striking armoured fish, such as *Dunkleosteus*, were common at this time, along with sharks and the first ray-finned fish. The first ammonites also appeared, floating through the sea on a planet still dominated by a single great ocean. On land, however, arthropods began taking over fresh habitats, which were being radically altered by plants evolving true roots and seeds. Venturing to feast on this bounty were the earliest four-limbed vertebrate animals, the tetrapods, taking their first steps tentatively onto land. A mass extinction at the end of the Devonian reshaped the world, snuffing out organisms that had dominated our planet for millions of years, and setting the scene for the rise of tetrapods.

In the Devonian, our planet was still dominated by the massive Panthalassic Ocean. This expanse of blue spanned most of the northern hemisphere and much of the equatorial regions. Gondwana still lay in the south, an expansive continent surrounded by smaller scatterings of land. Reef-fringed islands reached northwards across the equator. In the Devonian these little pieces of earth drew closer together, beginning to coalesce into new large landmasses, such as Euramerica. Sea levels were quite high, creating copious shallow ocean environments with bustling reef ecosystems. The climate was relatively warm globally, with hot conditions at the equator and a more temperate climate in Gondwana itself.

The most obvious evolutionary drama taking place in the Devonian oceans was among fishes. Sharks proliferated into countless new groups, as did ray-finned fishes. By far the most iconic piscene residents were the armour-plated fishes, or placoderms, whose bodies fill the world's Devonian rocks like thousands of discarded shields after a battle. They shared their ocean with trilobites, which were still common, and one of the other most iconic fossils in the world, the ammonites – or more correctly, ammonoids – which first appeared at this time.

This time period was also an incredibly significant moment for life on land. Plants grew in size and formed the first extensive forests. This was possible because of the appearance of true roots and leaves, and by the end of the Devonian, the first seed-bearing plants evolved, called gymnosperms. Horsetails and ferns also appeared, and together this verdant growth would have global consequences, creating a carbon sink. By sucking up CO_2 from the atmosphere, they caused the climate to cool. Changing global climate probably played a significant role in the Late Devonian mass extinction, in which shallow marine organisms and reefs were hit hard.

Despite this turbulence, something incredible emerged from the weed-choked pools of the Late Devonian world. A few bony, jawed fish were sticking a tentative tiptoe out of the water. Vertebrates were finally joining the many arthropods already established on land, including the ancestors of scorpions, millipedes and spiders. For the first time, the composition of terrestrial life was taking on a structure we might recognize today, and our most ancient ancestors were the last ones to join the party.

Old Red Sandstone

The Devonian is famous for a particular group of rocks called the Old Red Sandstone. These layers can be found along the eastern coast of North America, and in Greenland, the British Isles, Ireland and Norway. The Old Red Sandstone played an important role in the early science of palaeontology, revealing environmental conditions in the Devonian, and yielding spectacular fish, arthropod and plant fossils from the beds of ancient lakes and rivers.

In Scotland, there are places where the Old Red Sandstone sits at a strange angle against much older rock layers. One of the most famous is at Siccar Point, where it lies at a right-angle to Silurian-age rocks. This feature is known by geologists as an unconformity. Seeing how these layers sat at odds with one another helped the Scottish geologist James Hutton understand how ancient the Earth was, and that geological processes can tilt entire beds of rock over millions of years.

Reef's Hit Hard

The mass extinction at the end of the Devonian had an enormous impact, but less on the terrestrial world than on the marine. Shallow, warm-water organisms such as corals suffered the heaviest losses, with the populations of many key organisms collapsing as the devastation rippled through the ecosystem.

The causes of the extinction are difficult to determine because it took place over such a protracted period of time. The development of forests, and increased weathering of silica-rich rocks altered the atmosphere, reducing carbon dioxide levels and cooling the planet. The booming life on land may also have increased the amount of soil and nutrients flowing from streams and rivers into the sea, creating algal blooms in the water column. Such blooms in the modern world can smother reefs, cutting out sunlight and reducing oxygen levels in the seawater, killing everything trapped below.

There is evidence for widespread anoxia – a lack of oxygen – in sediments from the Devonian. This could have caused deaths throughout the marine ecosystem. It also meant that as their bodies sank to the seafloor, they didn't decay as quickly and their remains were preserved. Over time, many of these rich organic layers were converted to oil; the rock layers deposited above them acting like a gargantuan wine-press, compressing and heating the organic material. Today this oil remains a major source for human industry in places such as North America.

Dunkleosteus – Age of Fishes

The most recognizable of all Devonian inhabitants is *Dunkleosteus*, a gargantuan armour-plated fish with blade-like jaws. Some species grew as long as a bus, and were undoubtedly the scourge of the marine world. This was a period in which jawed vertebrates spread into and took over the world's seas and lakes, changing the entire course of vertebrate evolution.

Dunkleosteus was a placoderm, or 'armoured' fish, that lived over 360 million years ago. This enormous predator looked like a cross between a shark and a can-opener, with a terrifying mouth rimmed with notched, blade-like jaws. There are around ten species of *Dunkleosteus*, including some of the largest armoured fish of all time. The most infamous is *D. terrelli*, at over 7 metres (23 feet) long. *Dunkleosteus* fossils are found in North America, Europe and Northern Africa. This top-tier predator probably swam relatively slowly, but it had a lightning-fast, bone-shattering bite.

As their name suggests, armoured fish wore an outer coating of bony plates around their heads and 'shoulders', with the rest of the body wrapped in a chainmail of scales. This armour was arranged so that they could still move and feed with ease. Rather than teeth as we would recognize them, most placoderms had sharp, beak-like mouths, perfect for slicing, puncturing and crushing. These fish appeared in the preceding Silurian Period, and include the first known jawed fish, no bigger than a paperback book and called *Entelognathus primordialis*, which means 'primordial complete jaw'.

The development of jaws changed the course of evolution. They probably developed from gill arches: the struts that support the gills and help fish to breathe. The gill arches lying closest to the head fused and moved forwards under the skull, eventually forming the upper and lower jaws, as well as part of the braincase. Using their solid jaws, fish were able to grip food, manipulate it, bite or crush it. Some fish used their jaws for suction-feeding or jaw-protrusion – when they shoot their jaws forwards like a grabbing hand to snatch their prey. The emergence of jaws had knock-on effects for the entire marine ecosystem. Prey with hard exoskeletons were no longer safe in their shells. Animals had to become faster and develop defences such as spikes if they hoped to survive the hungry jaws that powered through the Earth's rich oceans.

The armoured placoderm fish, *Dunkleosteus*, hunting its prey.

74

Many Fish in the Sea

Placoderms such as *Dunkleosteus* were the most prolific fossil vertebrates of the Devonian, but they were far from the only ones festooning these ancient seas. Alongside them were the earliest sharks, creatures such as the agile *Cladoselache*, found in North America. It had a streamlined body and a softer cartilaginous skeleton, superficially resembling sharks as we know them today.

Meanwhile, bony fish were also diversifying. These animals had endochondral bone – which has a sturdier structure than other bone types – as well as tooth-bearing jaws, a distinct skull and body scales. The bony fish form two main branches of life: ray-finned and lobe-finned fish, which together have more species than any other vertebrate group on Earth. Four-limbed animals, the tetrapods, evolved from lobe-finned fish. You might even say we are just extremely unusual land-living bony fish.

Scottish Sex Lakes

Some of the most famous Devonian fossil-bearing beds in the world are found in Scotland. The animals within them tell saucy tales about sex in the age of fish. The richest rock layers were once the bed of a semitropical freshwater lake, named Lake Orcadie. It lay in what is now northern Scotland and the North Sea, and once bustled with crustaceans and molluscs, fed upon by many types of fish, including placoderms. As water levels rose and fell through time, this habitat periodically shrank and even dried out, causing mass die-offs in the animal population. These moments in deep time left behind 'fish beds', comprising literally hundreds of fossil fish piled on top of one another like autumn leaves.

Devonian fossil fish beds were crucial to our early scientific understanding of the fossil record, and they continue to yield scientific marvels today. A cousin of *Dunkleosteus*, a fellow placoderm named *Microbrachius* recently provided the first evidence of copulation in the fossil record. Fish vary in how they fertilize their eggs: some fertilize internally, others spawn – releasing eggs and sperm to mix and fertilize outside the body. It turns out that *Microbrachius* mated side-to-side, linking its front fins together as the male transferred sperm to the female via L-shaped genital claspers. At 385 million years old, this is among the oldest evidence for internal fertilization in vertebrates. It is thought that internal and external fertilization have evolved multiple times in the evolution of fish – including of course, in the ancestors of tetrapods.

76

The results of this sexual 'line dance' are also preserved in Lake Orcadie's sediments. Embryos were found in the belly of pregnant placoderm called *Watsonosteus*, which lived and died around the same time as *Microbrachius*. Their tiny underdeveloped bones constitute the oldest vertebrate embryos yet known in the fossil record.

Apex Predators

The never-ending interplay between predator and prey began with the first complex life on Earth. In the study of ecology, complex food webs trace the movement of energy through an ecosystem. At the 'top' is the apex predator, an animal that eats others and has no natural predators. Although we might like to think of ourselves this way, humans are not apex predators. We have a mixed diet from multiple trophic levels, and despite all our technology, we have many natural predators. Apex predators fascinate humankind. When it comes to the fossil record, we are especially attracted to large-bodied carnivorous animals such as *Dunkleosteus*.

Since our earliest origins, our relationship with such animals has never been easy, resulting in their reduction and extinction across many continents. Their disappearance has highlighted their crucial role within ecosystems, where they are often keystone species (those that have a profound importance to the healthy functioning of an ecosystem). Predators control prey numbers, which in turn has an impact on the grazing of plants. In an example from Yellowstone National Park in the United States, the reintroduction of wolves (previously hunted to extinction by humans) changed the numbers and feeding behaviour of herbivores, allowing over-grazed plant species to recover. This regenerated entire habitats for other species to occupy, and the diversity of the whole park was revitalized.

Dunkleosteus and other placoderms were among the many casualties of the end-Devonian mass extinction. However, there were plenty of groups waiting in the wings to replace them. Sharks in particular have come to excel in the marine predator role, but through geological time many animal groups have taken on this niche in the sea and on land. These creatures ensure the vital functioning of ecosystems over millions of years, and leave their dramatic remains throughout Earth's fossil record.

Pneumodesmus – First Animals on Land

From the end of the Silurian and into the Devonian, the continents of our planet became home to the first land-living animals. The millipede-like *Pneumodesmus* was among the first of these arthropod pioneers. Alongside it, the early relatives of spiders and scorpions took advantage of the plant and fungal ecosystems that had already established themselves on the Earth's surface. By feeding, breeding and dying on land, they added new tiers of complexity to the terrestrial food web, and provided a reward to the others who would later venture from the water's edge.

With the surface of the Earth having turned green with plant life, it was only a matter of time before animals followed them ashore. Not long after the first vascular plants were making their home in the young soils of our continents, arthropods set foot in the undergrowth. Among the oldest evidence for these intrepid explorers is a fossil found near Aberdeen in Scotland, named *Pneumodesmus*. It is a myriapod, the group that includes millipedes and centipedes. Although originally dated to around 423 million years ago in the Silurian, recent studies suggest it may be younger, living in the earliest Devonian. Whatever the case, by the Devonian, animals had firmly established themselves out of water, and *Pneumodesmus* is among the very earliest to walk on our planet.

Pneumodesmus may have resembled this modern millipede.

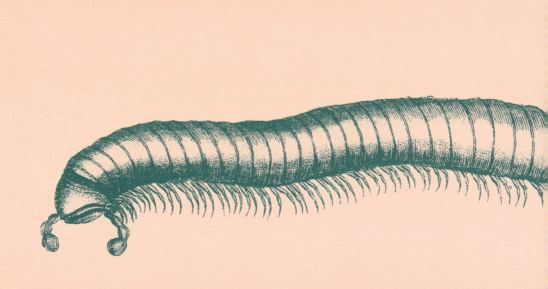

Only one fossil of *Pneumodesmus* exists, and it is only a 1-centimetre (2/5-inch) piece of the body. But in that tiny section, multiple legs are visible wiggling from six segments of a recognizably millipede-like animal. More importantly, details of breathing structures are visible: holes in the cuticle of its exoskeleton, called spiracles. These allow oxygen and other gases to enter and leave the body, and it was this feature that gave the fossil its name, *pneumo*, from the Greek for 'breath' or 'air'. This fossil provides the first definitive proof of air-breathing, a radical new evolutionary adaptation that opened up the surface of the continents to millions of tiny arthropod explorers – and the predators who followed them.

Arthropod Earth

In the Devonian, *Pneumodesmus* was not living alone among the vegetation. Alongside it were many other myriapods, their oldest fossils coming from Silurian and Devonian rocks. Although they don't fit into any modern groups of millipedes or centipedes, they are early relatives of those alive today and superficially resemble them, possessing long segmented bodies and many legs – millipedes have two legs on each side per body segment, whereas centipedes have one. The millipede with the most legs known today is *Illacme plenipes*, which has up to 750 legs. Most millipedes today are detritivores,

feeding on decaying plant matter. The fossil record for these animals is sparse, making each one especially precious for our understanding of life's emergence from water. It is likely the first myriapods were attracted onto land by the new food source generated by early plants.

The first arachnids were also taking advantage of the world opening up above them. This is the group that includes mites, scorpions, spiders and harvestmen. They have eight legs (unlike insects, which have six), and most of them still live on land, although a few (such as the diving bell spider, *Argyroneta*) have returned to life in water. Fossils from the Ordovician and Silurian suggest arachnids and other arthropods may have occasionally come out onto land much earlier, but by the Devonian, some had fully transitioned to air-breathing. These first arachnids were trigonotarbids, an extinct group that looked like a cross between a spider and a mite. Mites and pseudoscorpions were also abundant, and later, spider-like animals such as *Attercops*, which had spinnerets capable of producing silk. Just like today, most of these early arachnids were predatory, probably feeding on the other arthropods venturing from the water's edge.

By the end of the Devonian, these animals are joined by the first insects, which comprise an estimated 90 per cent of all animal life on our planet today. Finally, some vertebrates also made the transition to land, perhaps driven by the search for new food sources. The underpinnings of terrestrial life as we know it were finally in place. From there, evolution has worked on these groups to create the incredible diversity and abundance we see around us today.

What Are They Good For?

Arthropods are often seen as pests, particularly the insects. However, they play a fundamental role in the functioning of the entire planet. There are now over 16,000 myriapod species, 60,000 arachnids and an estimated 10 million insect species. Not only were they important in the earliest ecosystems on Earth, but they also continue to be vital both to the natural world and the human one.

Myriapods process leaf litter in forests, making them a crucial cog in the nutrient cycle. Centipedes are usually predatory, with the largest capable of eating small mammals and reptiles. Arachnids are also mostly predatory, and so play their part in regulating the populations of their prey. These include insect pests which, in uncontrolled numbers, would otherwise cause damage to plant pop-ulations. As a result, the humble spider is vital to human agriculture. Mites and ticks can be parasitic and carry disease, posing a threat to humans and other animals, and some other insects pose similar dangers. Yet the roles of insects are so varied that their value is truly incalculable. This includes those that produce honey and even engineer entire ecosystems with their industrious activity, such as bees, ants and termites.

Many arthropods are venomous, including those deadly to people. However, venoms that disable and kill their prey can also be beneficial; spider venom has been used as an alternative pesticide and is being investigated for use in medicine and new materials. Arthropods can also provide a food source for countless animals, including one another. Many are eaten by humans, including tarantu-las and scorpions, grasshoppers, termites and weevils. As many as 2,086 species are eaten around the world today, and have been a source of food since at least the Palaeolithic. It is suggested that insects, in particular, may provide an important source of protein in the future as human populations continue to increase – an alternative to resource-intensive meat rearing.

It is hard to imagine our planet without arthropods; indeed, it probably couldn't exist. Back in the Devonian, the world was theirs for the taking. But where they ventured, their predators were not far behind. The presence of arthropods provided food for another group of animals emerging from the water, and one that is particularly important in the evolutionary history of humans: the tetrapods.

Ammonites – Spirals of Geological Time

Ammonites are one of the most common and recognizable fossils. These marine molluscs lived in the sea, growing inside spiral shells where their soft bodies were protected. They had multiple squid-like arms protruding from the open end, searching for food. After thriving for over 340 million years, they became extinct around the same time as the non-bird dinosaurs. As well as being an important index fossil for geologists, ammonites have also played a major role in folklore for centuries.

During their epic 340 million-year existence, ammonites (or more correctly, ammonoids, the group to which ammonites belong) populated seas and oceans all over the world. They came in a mind-boggling range of sizes: the very smallest would fit easily on your fingernail, whereas the largest were over 2 metres (6 feet) across. They are a type of mollusc, the group that includes slugs, snails and squid. Their soft, squishy bodies are rarely preserved, so it is difficult to work out the details of their anatomy and way of life. It is likely that most of them lived in open water, from shallow seas to the deepest oceans, but probably not in brackish or freshwater environments. Some species fed on plankton, and there is evidence that they may have squirted ink for defence, similar to their relatives, squid and octopuses.

Despite their incredible success, ammonites died out at the end of the Cretaceous in the same mass extinction that killed the non-bird dinosaurs. Only their close relative, the nautilus, is alive today. Their sudden disappearance remains a mystery, but it is likely that the mass extinction decimated plankton in the oceans. This plankton included both the ammonites' primary food source and their eggs, leaving them unable to recover. However, ammonite fossils are so commonly found and easily recognized that their legacy lives on, incorporated in human cultures around the world. Their spiral ghosts are an enduring symbol for palaeontology among scientists and the general public alike.

Ammonite shells are made up of multiple chambers that increase in size from the centre of their spiral shell outwards. Most of these were empty, with the animal itself occupying only the largest chamber at the opening of the shell. The rest of the chambers were likely filled with gases that could be adjusted to allow the animal to float through the water column. The ammonite's soft body had as many as ten arms,

A selection of ammonite shells, showing their diverse shapes and sometimes open coils.

digestive organs, probably an ink sac and gills. Many of them had a hard palate in their mouths called a radula, which allowed them to crush plankton or larger prey.

Although we associate ammonites with their trademark spiral shape, there were species with very different shell structures. These irregularly coiled ammonites are collectively called heteromorphs. *Scaphites*, for example, grew in the shape of a number '9', whereas *Baculites* was almost completely straight, resembling a longbow. The bizarre ammonite *Nipponites* looked like a pile of discarded string, growing in all directions.

Ammonite Zones

Ammonites existed for a huge portion of Earth's history. There were thousands of species that are relatively easily identified by the shape of their shells and the sutures between their shell chambers. Species often originated and became extinct over very short time periods, some species persisting for as little as 200,000 years. This, coupled with their global distribution in the oceans and seas, makes them an ideal index fossil.

Hundreds of ammonite 'zones' have been established by geologists and can be used to correlate the age of rocks. This approach to understanding rock stratigraphy (the study and classification of rocks) was pioneered in the 1850s by German geologists Friedrich Quenstedt (1809–89) and Albert Oppel (1831–65), based on their research in the Jura Mountains of France and Switzerland. Thanks to this work, pockets of rock can be compared across huge distances, using the species preserved within to work out their age and context within the yawning depths of geological time.

Ram's Horns

Ammonites are so widespread that their fossils have a rich cultural legacy, recognized and collected by people around the world for millennia. The name ammonite comes from the Egyptian god Ammon, who had ram's horns resembling the shape of this fossil mollusc. In China ammonites are also likened to ram's horns, and in Chinese folklore they represent an unknown animal that had been turned to stone. Ammonites have been part of human cultural beliefs and practices since at least the Mesolithic. There is an ammonite incorporated into one of the upright stones of a Neolithic long barrow near Bath in England. Ancient Sanskrit texts suggested that ammonites were the traces created by some kind of worm. In Hindu traditions they are considered precious for their resemblance to the chakra held in the hands of the God Vishnu.

As well as inspiring origin stories, these fossils have also been used for medicine and ritual. The Ancient Greeks believed ammonites cured blindness and infertility, and protected people from snake bites, whereas the Romans placed them under their pillows to give prophetic dreams. In medieval Europe ammonites were thought to be the remains of snakes that had been turned to stone, often by saints. These 'snakestones' were used as charms to heal farm animals, or cure ailments such as bites and stings. In North America ammonites are called 'buffalo stones' by the Blackfoot people, and used in hunting ceremonies. The Plains and Navajo Nations called them *wanisunga*, 'life within the seed, seed within the shell'.

Acanthostega – First Vertebrates on Land

At the end of the Devonian, the first vertebrate animals ventured onto land. From the weed-choked waterways of humid equatorial regions, members of the lobe-finned fish lineage emerged from their swampy world. One of these pioneers was *Acanthostega*, a salamander-like creature that lived in what is now Greenland. Although not yet fully land-living, its body provided the scaffold for evolution to work upon, adapting to the many challenges of life out of water.

Animals such as *Acanthostega* were among the predecessors of all land-living vertebrates. This early tetrapod lived in the late Devonian, around 365 million years ago, and is known from fossils found in Greenland. It lived in shallow, swampy environments, and had both gills and lungs. It would have been just over half a metre (20 inches) long and had a flat, wide head and four limbs, with a long oar-like tail perfect for movement in water. *Acanthostega* resembled the modern giant salamanders found today in the eastern United States, China and Japan, but it was no amphibian – although it was the predecessor of that group, as well as of reptiles and mammals. This creature was related to the ancestors of all land-living four-limbed animals on Earth.

Despite looking ready to walk out of the water and into the next chapter of the evolution of life, we know from the shape of *Acanthostega*'s shoulder bones that it was unlikely that it could have actually walked on land. However, the limb architecture was in place for later tetrapods to develop upon. There are now a number of animals like *Acanthostega* known in the fossil record, particularly from Arctic Canada and Greenland, places which once lay near the equator. As a group, they are called early (or stem) tetrapods. These include more fish-like animals such as *Panderichthys* and *Tiktaalik*, which had less well-developed limbs. These animals evolved from among the lobe-finned sarcopterygian fishes, which were abundant in the Devonian but today are only represented by the lungfishes and coelacanths. *Acanthostega* is one of the first early tetrapods to have a well-developed pelvis, giving it more powerful back legs. This change made it possible for later tetrapods to leave the water completely.

Acanthostega (below) and *Ichthyostega* (above), some of the first tetrapods to walk on land.

**Paddling
Before Walking**
_____

It is unlikely that the first early tetrapods could walk on land – most were primarily aquatic animals. Despite the misconception that legs evolved 'to allow' animals to walk, this phraseology approaches the process of evolution back-to-front. Major evolutionary adaptations – such as those that made it possible to walk on land – usually come from a suite of anatomical adaptations that are driven by other factors, such as predation or environmental change. Such modifications don't occur to achieve a goal but are selected in response to external stimuli.

The first tetrapod limbs probably evolved as sarcopterygian fish used their fins for movement underwater. These fins may have grown larger and bonier as they were increasingly employed to help the animal push its way through weed-choked lakes. Animals such as *Acanthostega* didn't have wrists, and their limbs projected outwards laterally from the body, like a falling skydiver. This meant they were unable to bring their feet underneath their bodies to support their weight, but they were able to crawl and pull themselves through the tangle of underwater weeds. Many of the earliest tetrapods also experimented with the number of digits on their hands and feet. *Acanthostega* had eight fingers and toes, but later, five would become the established number for the tetrapod body plan. This was probably the optimal number of digits to support weight while still allowing free movement in the wrist and ankle.

However, there is evidence to suggest these early, mostly aquatic tetrapods were already spending some time out of water. Their teeth were different from those of other fish, and the architecture of their skulls meant they were able to bite, rather than using suction or other types of feeding. This suggests they may have eaten prey from the water's edge, such as arthropods, bringing their heads above the surface to grab them. *Acanthostega*'s pelvis was also fused with part of its vertebrae, making it more stable and perhaps able to bear some weight for short forays onto the surface.

**Sticking Your
Neck Out**
_____

Although people often focus on the limbs of the first tetrapods when talking about their transition onto land, the development of the lungs and necks may have been just as – if not more – important. The first lungs probably developed in response to periodically low oxygen levels in the shallow pools where these animals lived. As lungs evolved, it meant that early tetrapods could take advantage of there being up to thirty times as much available oxygen in the atmosphere as there is in water. Animals such as *Acanthostega* breathed using a kind of bellows system, which drew air into their wide mouths and pushed it down into their lungs by lifting and dropping the underside of the mouth.

Later, tetrapods evolved a new way to breathe. They started using their chest muscles to inflate and deflate their lungs, a system known as costal breathing. No longer needed for respiration, their heads became narrower in shape. These animals developed the first necks, holding the skull further from the rest of the body. This meant that parts of their skull and jaw muscles could be repurposed for feeding, giving tetrapods a stronger and more precise bite. This was an important development, as it opened up new methods of feeding and changed their body plan permanently. Finally able to venture from the water's edge, tetrapods continued to adapt, and became an integral part of the terrestrial ecosystem. Life on land would never be the same again.

Carboniferous

The hot and heady world of the Carboniferous began 359 million years ago and lasted 60 million years. Despite its vintage, it is one of the most important periods for humankind, both in terms of our evolutionary story and because it fuelled our industrial present. This period saw the rise of expansive swamp forests, coating a greenhouse-planet dominated by giant insects. But dramatic climate change soon redrew the face of the Earth, paving a path for our earliest four-legged ancestors and creating ever more complex food webs on land.

For the first time in the Carboniferous, our continents took on the technicolour topography of a familiar world. The glittering Panthalassic Ocean made a blue marble of most of our planet, while the land glowed green with verdant forests. For the whole time period, glaciers formed a white cap on the southern pole, over which much of the great continent of Gondwana was still resting. Some ice may also have covered the North Pole and the tips of the landmasses, which continued their drift northwards. What is now Siberia and Kazakhstan lay at the highest northern latitudes, with Euramerica and the fragmented components of what would later become China positioned below them, near the equator. These bits and pieces of the world were colliding throughout the Carboniferous, joining to form a new landmass, bejewelled with fresh mountain chains. By the end of the Carboniferous, the creation of the supercontinent of Pangaea was almost complete.

For most of the Carboniferous, the Earth was predominantly warm and lush. Pristine swamp forests coated much of the landscape, giving it a lush terrestrial ecosystem swarming with insects. The highest atmospheric oxygen content in our planet's history (35 per cent, versus 21 per cent today) allowed these insects to grow immense, including the largest arthropods to ever walk the Earth. Millipedes grew longer than cars, and dragonflies reached the size of seagulls. The earliest tetrapods also thrived in the swamp forests. Their fossils have been found preserved inside tree trunks, where they probably took shelter or hunted insect prey.

Sea life continued to prosper, with diverse reef ecosystems picked over by roaming ammonoids and myriad fish. Trilobites, the stalwarts of the ocean since Cambrian times, became rarer. The brachiopods, shelled animals resembling molluscs which are rare now, were abundant and diverse in the Palaeozoic, alongside giant sea scorpions and crustaceans – including the ancestors of crabs. As the armoured placoderm fish declined, sharks cruised into their empty niches. There were shell-crushing specialists, and species with circular arrays of razor teeth. Not all of their anatomy is understood – for example, *Stethacanthus* had a flattened disc projecting from its back like an ironing board covered in spikes, the purpose of which remains unclear. For all that the world was becoming more familiar, it was still a strange and alien place.

Romer's Gap

For many years there was a conspicuous gap in the fossil record at the start of the Carboniferous Period. It was identified by the palaeontologist Alfred Sherwood Romer (1894–1973), and later named after him. This 15 million-year time period yielded few fossils globally, but clearly marked a significant moment in tetrapod evolution. Before the gap, early four-limbed animals such as *Acanthostega* were barely able to leave the water, but afterwards they were comfortably adapted to a terrestrial life. Without fossils from the intervening period, it was difficult to understand how this transition occurred.

For years people tried to determine the cause of Romer's Gap, suggesting that environmental conditions may have hindered the formation of fossils or caused an ecosystem collapse. But recent research suggests that the lack of fossils says more about human industry than the ancient world. It is likely that the earliest Carboniferous rocks simply haven't been explored for fossils as thoroughly, because there weren't coal seams and other industrial resources in them to garner the interest of geologists. As more sampling is carried out, it appears that there is not such a conspicuous 'gap' in life forms as previously proposed. New discoveries are currently being made that are filling in this part of the evolutionary story.

Black Gold

The largest coal deposits in the world date from the Carboniferous, giving the time period its name. As the first forests grew, they absorbed and locked atmospheric carbon into their tissues. When these plants died they formed deep layers of rotting vegetation, and these became the coal seams that riddle our rocks. But disaster struck these first forest ecosystems: around two-thirds of the way through the Carboniferous the climate cooled and dried out, devastating the species that had thrived in the hot swampy world. The mass disappearance of the swamp forests is called the 'Carboniferous Rainforest Collapse'.

Coal is formed from dead plant material that first decomposes into peat, and is then buried in sediments. Over millions of years, this burial creates intense heat and pressure, which drives out the water, carbon dioxide and methane, leaving a high proportion of carbon behind. As this process continues, the plant material is transformed from peat to lignite (also called brown coal), then bituminous coal and finally anthracite, or black coal. Jet, a black gemstone found around the world, is a type of lignite that has been prized for carving jewellery and ornamentation for thousands of years.

As we burn fossil fuels, we pull out the carbon long trapped by the Earth and release it to cause havoc in the atmosphere. The very life that sustained the formation of our world has now become the source of human-induced climate change, heralding an uncertain future for evolution on our planet.

Lepidodendron – First Forests

The lush swamp forests of the Carboniferous are an emblem of this time period. As well as providing the thick coal seams that fuelled human industrialization, they are sometimes preserved as 'fossil forests'; sites where their ghostly trunks remain upright in situ, as though recently felled. However, appearances can be deceptive – these 'trees' are in fact giant relatives of today's diminutive understory inhabitants. Their ancient relatives provided new habitats on a world bustling with the first land-living organisms. They were replaced by drier woodland as the Earth's climate changed, leaving them forever in the forest shadows.

The first forests of Earth emerged in the Devonian, but by around 350 million years ago, they had spread to coat the warm, oxygen-rich world of the Carboniferous. These trees were unlike those we know on Earth today. They were dominated by huge relatives of the lycopsids, such as quillworts and clubmosses, plants which still exist today but are mostly less than 20 centimetres (8 inches) in height. These ancient predecessors are also known as scale trees, and they include *Lepidodendron* and *Sigillaria*. They grew over 30 metres (98 feet) in height, towering over a dense understory of horsetails, ferns and mosses. Their fossils are found around the world, representing a time when the swamp forests stretched from shore to shore across the ancient continents.

Although tree-like in stature, it seems *Lepidodendron* grew quite differently. Thanks to the tropical climate of the Early Carboniferous Earth, it could reach maturity swiftly. It is thought *Lepidodendron* may have only lived for only around 15 years before reproducing and dying. The trunk, which was up to 2 metres (7 feet) across when fully grown, was covered in needle-like leaves that resembled pine needles. These dropped off as it grew, leaving a pattern of knobbles like plucked chicken skin. The trunk was long and straight, with no branches until the crown of the tree, which held the remaining leaves and cones.

Ancient lycopsids such as *Lepidodendron* covered a huge part of the terrestrial world in the first part of the Carboniferous, spanning 120 degrees of latitude. Their reign was not to last: as the climate shifted in the latter half of the time period, the land dried out and became unsuitable for them. Although *Lepidodendron* could still be found in some parts of the world, it was never again the dominant organism in the forest. It finally became extinct at the end of the Triassic, leaving only its diminutive cousins behind.

Lepidodendron formed vast swampy forests around the world. They are related to quillworts and clubmosses.

92

Known by Many Names

The way palaeobotanists – scientists who study fossil plants – name their specimens is quite different from other extinct groups. Whereas most organisms are given a binomial genus and species name unique to that organism, extinct plants can be known by multiple names. This is because plant fossils are so often found as fragments, and so the different parts of extinct plants are named separately. Like scattered jigsaw pieces, it is often years or decades later that the plant is finally pieced together and the whole picture revealed. *Lepidodendron* is also known by multiple names. One of the most common is *Stigmaria*, which are the underground rooting structures of the tree. Other names that may be attributed to the same plant are *Bergeria*, *Knorria* and *Aspidiaria*.

Fossil Forests

Lepidodendron and other Carboniferous plant fossils were popular in natural history collections and art in 19th-century Europe and beyond. Due to the pattern of their bark, they were often displayed in fairgrounds and amateur exhibitions as examples of ancient fossil lizard or snake skin. Most scientists, however, recognized that they belonged to extinct forms of plant that were also the source of coal. Due to the pursuit of coal for industrialization, more plant fossils were unearthed than ever before, painting a picture of a richly botanical ancient world.

Some of the most stunning and prized fossils were those of multiple trees preserved together upright and in situ, like a small section of felled forest. Examples such as the Fossil Grove in Glasgow, Scotland, and in Saint Etienne, France, represent the infilled casts of the base of *Lepidodendron* trunks and their roots. They helped palaeontologists understand processes of fossilization such as rapid burial, and how this could result in three-dimensionally preserved fossils and rock casts that last millions of years through deep time. They also told them that the climate must have been very different in the past to support these ecosystems. Such coal-field fossils were a vital lesson in the study of geology, and shaped our understanding of the changing face of our planet.

Rainforest Collapse

In the latter part of the Carboniferous there was a complete switch-over in plant floras. The *Lepidodendron* forests had grown in the warm, moist environment of the Early Carboniferous, but by 305 million years ago, the climate started to dry out. This led to the fragmentation of the swamp forests, before finally causing their collapse across most of the world. Only a handful of refugia remained. The new habitats that replaced the swamps included forests dominated by tree ferns, and a group called the gymnosperms. This includes the relatives and predecessors of groups we know today, such as conifers, cycads and ginkgoes. Although widespread, these and the many other tree species that evolved afterwards were rarely preserved as coal in the same depths and thickness as the layers of scale trees of the Carboniferous.

Why did swamp forests preserve so differently from later habitats? The Carboniferous scale trees such as *Lepidodendron* had much higher ratios of bark to wood than modern tree groups. This bark not only supported the trees, but also offered protection from insect damage and forest fires, which were frequent due to the higher levels of oxygen in the atmosphere. One theory is that because this thick bark was composed of up to 60 per cent insoluble lignin, it was hard for organisms to break down and decompose it. This would have meant that for millions of years, the short-lived *Lepidodendron* and other plants didn't rot very quickly after they died, and so their trunks piled up to create deep organic layers, which became the thick coal seams that fuelled our Industrial Revolution. However, recent research suggests that it was more to do with the ever-wet tropical conditions of the forests themselves, which prevented rotting. Either way, towards the end of the Carboniferous conditions changed, reducing coal formation from that point onwards.

As the Carboniferous gave way to the Permian, the climate dried out still further, putting an end to the expansive rainforest ecosystems. As the supercontinent of Pangaea coalesced, inland habitats became arid, and deserts soon swallowed swathes of the landscape.

Meganeura – First Flight and Giant Insects

The Carboniferous world swarmed with giant arthropods. Thanks to the high levels of atmospheric oxygen, insects and other invertebrates could reach colossal size, many larger than dogs. As well as growing large, they were also the first group of animals to take to the air. Combining these two impressive traits, *Meganeura* was a giant dragonfly relative with a wingspan as long as your arm. It flew over a land filled with many-legged behemoths.

Meganeura was a huge dragonfly-like insect that lived in the Carboniferous Period, around 300 million years ago. Although not a direct ancestor of modern dragonflies (Odonata), it is a close relative and looked very similar, with a long cigar-shaped body, two sets of wings up to 70 centimetres (28 inches) across and large eyes. Like modern dragonflies it was also a hunter, feeding on insects and other invertebrates. There are now around six thousand species of dragonfly and damselfly (their closest relatives) on our planet. They live the first part of their lives as wingless freshwater nymphs, before emerging in spring and summer to take flight from the pond's edges.

Giant arthropods were not restricted to the skies in the Carboniferous. Below, equally impressive invertebrates stalked the undergrowth. A type of myriapod (the group that includes millipedes and centipedes) called *Arthropleura* was the largest land invertebrate of all time, growing to 2.5 metres (8 feet) in length. With a body made up of around thirty segments and at least forty legs, it wound its way through the undergrowth eating plants and decaying matter. Fossil trackways (called *Diplichnites*) of this beast have been found in many places, but most famously in Scotland and in Nova Scotia, Canada.

Although *Meganeura* and *Arthropleura* were the largest of their kind, there were still plenty of small species living alongside them in the Carboniferous. By the end of the period, the earliest relatives of mayflies, dragonflies and cockroaches were thriving, along with the spiders and millipedes that preceded them. All of these made rich pickings for the proliferating tetrapods – the ancestors of all vertebrate life on land – which were branching into new groups and exploiting this nutritious source of food.

The largest insects today include the titan beetle, *Titanus giganteus*, which can reach the length of an adult human hand. Although

Meganeura was an ancient relative of dragonflies that grew as big as a hawk.

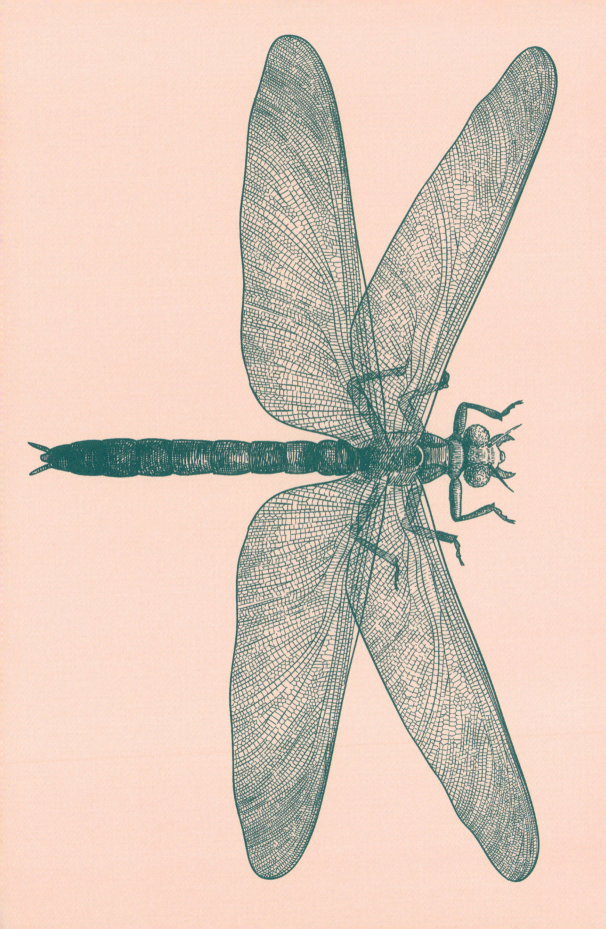

there are a few large species currently dwelling on Earth, of the many millions of insect species alive these are by far the minority. Most arthropods are extremely small by human standards, some are even microscopic. One of the main reasons that insects and other arthropods don't attain larger sizes regularly is that they are unable to absorb enough oxygen from the atmosphere.

Insects have a range of body plans, but something they do not possess are lungs. Instead, gases enter their bodies through holes called spiracles, often found on the thorax or abdomen. The earliest evidence for spiracles is found in *Pneumodesmus* (see page 78), a millipede-like creature that lived in the Silurian or Devonian. Gases enter the spiracles and circulate to the internal organs and tissues through a system of tubes called tracheae. As an insect grows larger, these holes and tubes become less efficient at gaseous exchange. This places a natural upper limit on insect body mass of around 100 grams (3½ ounces). In bigger insects the tubes become larger and more numerous, which increases the risk of the body drying out, resulting in death.

In the Carboniferous, however, the rules were different, because the atmosphere contained 14 per cent more oxygen. This raised the upper limit for body size, and insects and other invertebrates were able to grow gigantic. As the only flying organisms at the time, it is thought that the insects' lack of aerial predators meant creatures like *Meganeura* could become the flying apex-hunters, a situation that continued into the Permian.

First to Lift Off

The next time you swat a wasp or mosquito, it is worth remembering the incredible natural engineering that made it possible for them to buzz through our skies. Insects are the only invertebrates to evolve powered flight. They achieved this milestone in the Carboniferous, over 100 million years before the vertebrates followed suit in the form of pterosaurs, and much later, avian dinosaurs (birds) then mammals (bats).

It's not clear how flight first evolved, because there are currently no fossils to tell the beginning of the tale. Insect fossils are peppered throughout the coal seams of our planet, but in the Carboniferous they are mainly found in the latter half of the period. It is clear from these remains that flight had already emerged by that time, indicating that it likely originated much earlier. Wings probably developed from pre-existing parts of the insect body. One hypothesis is that they were modified from the paranotal lobes, structures that originally served as a kind of 'parachute' when the insects fell to the ground from great height, for example when evading predators. Another is that they are modified parts of their limb segments. With luck, future fossil discoveries from the Early Carboniferous might solve this great evolutionary puzzle and reveal exactly how these intrepid invertebrates took to the air.

Darting Jewels

Dragonflies play a rich role in human cultural history. Their life cycle, jewelled colours, beautifully architectural wing veins and supernatural speed when hunting have placed them firmly in myth and art for millennia. In the *Epic of Gilgamesh*, a poem from Mesopotamia that dates to over four thousand years ago, the change from water-bound nymph to flying dragonfly represented the impossibility of immortality. The dragonfly's speed symbolized industriousness to some indigenous Americans, and they were commonly depicted in pottery, rock art and necklaces made by the Hopi, Dakota and Pueblo.

In Japan, the dragonfly has inspired countless haiku and artworks, and one of the oldest texts from the country refers to Japan as Akitsushima, sometimes interpreted as 'Dragonfly Island'. The very colours and wing veins that make them so popular artistically are also used to identify between species. European folklore is less positive, however, painting dragonflies as injurious and sometimes even evil.

Pohlsepia – First Octopus

For thousands of years in human myth, many-armed monsters have arisen from the ocean depths to wreak havoc. But the origins of octopuses and squid lie much deeper, over 300 million years ago in the Carboniferous. *Pohlsepia* was a thumb-sized cephalopod that lived off the shore of North America's inland sea. From this tiny predecessor, octopuses have evolved extraordinary abilities for camouflage and problem solving, and their inky escapades have earned them recognition as the most intelligent invertebrates on Earth.

Today's squid (top left and centre) and octopus (top right and bottom) all descend from animals like *Pohlsepia*, that lived over 300 million years ago.

Pohlsepia was smaller than the top of your thumb. It had a round dumpling of a body with ten tiny arms – two of them shorter than the others. The remnants of what is thought to be an ink sac are visible as a stain in the centre of its fossil. Although hardly a terrifying denizen of the deep, this humble little invertebrate is the oldest known octopod on Earth. It foraged in the shallow nearshore waters of a large inland sea that covered part of North America 307 million years ago.

Octopuses and squid belong to a group of molluscs called cephalopods. These also include cuttlefishes and nautiluses, and the extinct ammonites and belemnites. The first cephalopods appeared as far back as the Cambrian, but *Pohlsepia*'s humble remains show us that octopuses and squid didn't emerge until almost 200 million years later. Although *Pohlsepia* is not a true octopus, it is certainly related to them. The lack of a shell is significant because it confirms *Pohlsepia* wasn't a member of the more numerous shelled cephalopods that shared its ocean home. Octopuses and squid both have eight arms, but only squid also possess two tentacles (with suckers on the ends) beside their mouth. The presence of *Pohlsepia*'s two tentacle-like appendages suggests that the shared ancestor of squid and octopuses had eight arms and two tentacles, and the latter were lost later in the octopus lineage.

Octopuses are unusual among cephalopods because they are exceptionally squishy, and as a result they rarely preserve as fossils. They possess a beak, or radula (the hard mouthparts that are a feature of all molluscs), but lack the cuttlebone found in cuttlefish and squid. Before *Pohlsepia* was discovered, the oldest octopus fossil was from the Jurassic. *Pohlsepia*'s discovery pushed their origins back another 140 million years, proving that these unique creatures have been part of our world for an incredibly long time. They now

inhabit every part of our oceans, from coral reefs to the darkest abyssal depths. Most are predatory, injecting paralysing saliva into their prey before pulling them apart with their powerful arms and radula.

Incredible Arms

Octopuses are remarkable. They can range from tiddlers to Leviathans such as *Enteroctopus*, the giant Pacific octopus that grows longer than a family car. Their sisters, the squid, can reach more than twice this length. Octopus skin is packed with cells that allow it to change colour for camouflage and communication. They have a complex, distributed nervous system that integrates the bombardment of information gathered by their tactile body. Their muscular bodies ripple with circular, longitudinal and transverse muscles that not only permit endless contortions, but coupled with their circular suckers makes it possible for them to hold fast to surfaces and prey, and manipulate objects.

There are chemoreceptors in the suckers of octopuses, and this means they taste everything they touch. They also have excellent eyesight, and are especially good at detecting movement. Most octopuses have an ink sac filled with melanin, a natural pigment. This can be shot out in an obfuscating cloud, allowing the animal to retreat from danger under cover of darkness.

The octopuses brain to body ratio is one of the highest for any invertebrate, close to the most intelligent mammals and birds. They are one of the only invertebrates known to use tools, such as holding together discarded coconut shells to create shelter. In aquariums, octopuses are experts at escape. Their skills have even been used to predict the results of sporting events. A German octopus named Paul predicted the outcome of football matches with a success rate of 86 per cent – although naturally, some have accused his aquarium keepers of cheating the odds.

Although there is no clear definition of intelligence, octopus escapades demonstrate an ingenuity unparalleled by most other life forms. Some scientists argue they have consciousness and feel emotion, albeit in a form so radically unlike our own that it has been hard to recognize. The likelihood that they can experience suffering led to a change in European law, giving octopuses, squid and cuttlefish – which are often used in scientific research – additional protections regarding their use in medical experiments. This makes them the only invertebrate animals legislated for in this way, acknowledging their unique status.

The Kraken Wakes

Octopus and squid are hugely significant to human culture as food, foe and fantasy. They are on the menu in countries around the world, but their maritime associations go beyond the culinary. Their sensuous arms can symbolize both eroticism and sinister danger. In reality, most octopuses are harmless to humans. Nonetheless, the hazards of ocean-going are often personified by these many-armed creatures, in the form of monsters that drag sailors to their submarine fate.

Octopus and squid are revered by the indigenous Ainu people of Japan's northern islands in the form of Akkorokamui, an unavoidable force of nature that can both harm and heal at whim. Octopuses also play a central role in the creation myths of some Pacific islanders. In Mediterranean Europe, octopuses have been depicted since the Bronze Age, and further north they reach throughout the sagas of Scandinavia in the form of the terrifying Kraken. In classic 19th-century French literature, octopus attacks have been used as allegories for the encroaching damage wreaked by the Industrial Revolution and scientific progress. The octopodian form with its many 'tentacles' has become a visual byword for horror, with their ability to stretch, contort, grab and penetrate.

In the long run, octopuses and squid probably have the edge over humankind. Not only have they persisted for over 300 million years, but their life cycle, intelligence and adaptability make them likely survivors of the current climate crisis. In some areas, certain octopuses and squid species are flourishing. As competitors are removed by overfishing and environmental damage, octopuses and squid have wider access to food. Their ability to reproduce quickly may make it easier to sustain numbers through difficult environmental conditions. Although most octopuses live only for a couple of years, females can lay as many as seventy thousand eggs at a time, which they often protect until hatching. Some studies indicate that rising temperatures are even accelerating their life cycles, although the picture of their response to climate change is far from clear. Ultimately, when we are gone, their descendants will likely remain as one of the most intelligent animals on Earth.

Hylonomus – Tetrapods Lay Eggs

The earliest ancestors of four-limbed animals evolved rapidly in the Carboniferous. From their precursors at the start of the period, they split into three major lineages: the ancestors of amphibians, reptiles and mammals. When climate change dried out their rainforest world, some of them adapted with a crucial innovation: the amniotic egg. This made it possible for them to populate the dry landscape of the ever-changing continents, taking over the end of the Palaeozoic with ferocious panache.

If you were to peer into the hollow of a *Lepidodendron* trunk in the Carboniferous, you might find inside it one of your oldest relatives. By around 310 million years ago, those first lobe-finned fish to venture landwards had evolved into fully terrestrial creatures. Among them were the oldest members of the mammal and reptile lineages, whose fossils have been discovered nestled in the ancient forests of our planet.

Hylonomus was a lizard-like animal not much longer than your hand. It lived in what is now Nova Scotia in Canada, and is one of the oldest undisputed members of the reptile line. It had sprawling legs, a long tail and a mouth full of conical sharp teeth. This was just one of many tetrapods (four-limbed animals) that scampered through the forests of the time period, feeding on the copious insects swarming in the high-oxygen atmosphere. Alongside it were the first members of the mammal line, called synapsids, such as *Archaeothyris*. Although superficially similar due to common ancestry, reptiles and synapsids are two completely distinct groups, and the details of their skeletal anatomy allows researchers to tell them apart. These reptile and mammal precursors lived alongside other types of early tetrapod, such as the ancestors of amphibians, as well as groups that later became extinct.

As changes in the planet's climate began to transform the landscape in the latter part of the Carboniferous, the ancestors of reptiles and mammals had a major survival advantage over their cousins. Unlike the other groups, *Hylonomus* and *Archaeothyris* were amniotes, so they laid shell-clad eggs. This protective coating and the membranes it contained meant they could reproduce even as their moist habitats dried, cooled and shrank to the margins of the continents. Soon the amniotes multiplied, giving rise to a newfound diversity of four-limbed backboned animals.

Hylonomus was among the first amniote tetrapods, laying eggs with protective shells.

Amphibians, Reptiles and Mammals

There's a common misconception that amphibians evolved first and that reptiles evolved from them, and mammals from reptiles. However, this is not correct. As more fossils have been found and our understanding of animal relationships has improved, we now know that although these three major lineages share common ancestors, they are totally separate from one another.

In the Early Carboniferous, the first tetrapods were *anamniotes*: they fertilized their eggs externally and laid them in water, as fish and amphibians do today. As their eggs are laid in water, anamniotes thrive in wetter habitats, hindering their spread to drier areas.

In the second half of the Carboniferous, the common ancestor of reptiles and mammals emerged. It was the first *amniote*, fertilizing its eggs inside the female and developing a more complex laid egg, enclosed in a leathery or hard shell. To our modern eyes, the first amniotes looked like the small reptiles that share our planet today, but this resemblance is only superficial. These ancient creatures were the distant forebears of both reptiles and mammals. By the end of the Carboniferous, fossils such as *Hylonomus* and *Archaeothyris* tell us that the amniotes had split into two main branches: the ancestors of reptiles (including birds), and the ancestors of mammals.

Although we like to simplify the story of life, in truth the early forests of Earth crawled with multiple strange and wonderful four-legged creatures. Working out how they are related to modern animals is challenging because the fossil record is sparse and difficult to interpret.

Egg Before Chicken

The split between anamniotes and amniotes is one of the fundamental divisions in the great vertebrate family. It would have enormous repercussions for land-dwelling vertebrates, moulding their evolution and success through geological time as the world unpredictably shifted, heated or cooled.

The most common terrestrial anamniotes today are the amphibians. Their eggs – known as spawn – are often laid in jelly bundles in ponds or streams. The developing embryo exchanges oxygen and waste directly into the surrounding water. Some species have developed different strategies to cope with low-water environments, including utilizing water collected in flowers, or even brooding their eggs in their mouths. However, most species still rely on water for reproduction.

For amniotes, the shell that encases the embryo provides its own portable 'pond'. Inside is a membrane that surrounds the developing young in its amniotic fluid, along with a yolk sac that feeds the embryo, and a structure that processes waste products. This radical new type of egg freed amniotes from ponds and streams, enabling them to move further inland and populate new habitats. It probably gave these groups an advantage in the later Carboniferous as the climate dried out, because their eggs were protected from desiccation, and could be buried to keep them at the right temperature for hatching. Over time, some amniotes – including marine reptiles and mammals – abandoned external hatching altogether, giving birth to live young.

Permian

The Permian was an otherworldly climax to the epic Palaeozoic. Beginning 299 million years ago, the Permian saw a Janus world of fluctuating climate and extreme environments. The continents finally collided to form the supercontinent of Pangaea, rimmed with monsoon forests and clasping a heart of hot dust. Conifers evolved, as did the first large tetrapod herbivores and carnivores. Yet this bountiful living world was not to last. Just 47 million years later, the Earth herself nearly wiped it all out, bringing the Permian to a brutal close that had monumental repercussions for the course of evolution.

The Permian was the last time period of the Palaeozoic and lasted from 299 to 252 million years ago. At this time, the Earth was a planet of extremes, one half water and the other half land. The Panthalassic Ocean lay unbroken from the eastern fringes of the shrinking Palaeo-Tethys Ocean to the planet-spanning western shore of a new supercontinent: Pangaea.

Pangaea had formed at the close of the Carboniferous. Throughout the Permian, it held newfound extremes of climate and landscape. Global temperatures varied a great deal through this time period, from the tail end of an ice age at the beginning of the Permian, followed by heating and drying in a series of warm and cool cycles. Pangaea's shores were lashed with monsoons that drenched glittering forests of conifers and cycads. As time went on, the heartland of the supercontinent desiccated, creating expanses of arid hills and deserts. These experienced intense high and low daily temperatures, challenging life to thrive. Trees with seed coverings did well in these new conditions compared to the moisture-craving ferns and lycopods. The last remnants of the swamp forests clung to islands around the fringes of the Palaeo-Tethys, in what is now southern China. The ancestors of many modern plant groups appeared, such as the conifers. A plant called *Glossopteris* became dominant in the southern reaches, leaving crucial evidence behind for geologists about the layout of our continents in the past.

Despite the desiccation, the Permian world was bursting with some of Earth's first modern food webs. The early relatives of cockroaches infested the landscape, and the first of the beetles and bugs (Coleoptera and Hemiptera) emerged. Tetrapods – especially the ancestors of mammals – grew large, and some adapted to feed on plants. The first herds of herbivores foraged the landscape, and exploiting them came large predators: nightmarish creatures the size of tigers, with sabre teeth. By the end of the Permian, these ancestors of mammals were flourishing. This familiar yet strange yin-and-yang planet was brought to an abrupt halt in the largest, most devastating mass extinction in history. It would reset the course of evolution, removing the mammal lineage from its hitherto main role in ecosystems, and instating reptiles as the new dominant vertebrate life on Earth.

Supercontinent Cycles

Supercontinents are formed when all of the dispersed major landmasses of our planet converge, like a geological rugby scrum. This cycle is driven by convection in the Earth's mantle – the transfer of heat through the molten innards of the planet – which rearranges the continental plates. Plates can collide like a slow-motion car crash, piling up into mountains, or plunging downwards to form deep trenches. They can also slide beneath one another, sinking into the depths and melting into the mantle. Volcanic eruptions often take place along these shifting plate margins.

Pangaea is perhaps the best known of the supercontinents that have patterned the face of the Earth. It formed at the close of the Carboniferous and lasted until the Early Jurassic. But it is not the only one; the cycle of drifting apart and reforming of land masses has occurred multiple times in the history of the planet. Not all geologists agree on the number and names of the previous supercontinents, but their formation and breaking up has occurred cyclically over the last 3.6 billion years, with as many as ten different supercontinents identified. The position of these continents affects wind patterns and ocean circulation, altering global climate. For this reason, their formation and subsequent break up has always had dramatic effects on environments, both on land and in the sea.

The Great Dying

It is no exaggeration to say that life nearly ended in the apocalypse that was the close of the Permian. As much as 85 per cent of species died out, including most of the major groups of tetrapods that had formed the planet's first megaherbivore–carnivore food chains. Among insects, which generally weather most mass extinctions without too much impact, entire taxonomic orders perished. These extinctions were not confined to the land: trilobites were the most iconic losses in the ocean, but species with calcium carbonate exoskeletons especially suffered, due to extensive ocean acidification.

The primary cause of the end-Permian mass extinction was a series of massive volcanic eruptions. A phenomenon called flood basalt eruption took place in what is now Siberia, smothering an area the size of Australia in lava. These lava flows formed the distinctive landscape known as the Siberian Traps. As well as killing everything it touched, the eruptions released enormous volumes of ash, sulphur-rich gases, methane and carbon dioxide into the atmosphere. These gases locked in heat, generating a runaway greenhouse effect that baked the planet. The chemicals reacted with rain to fall as sulphuric acid. Together, this decimated plant and marine life, disrupting food webs planet-wide, and obliterating much of animal life.

In the Triassic Period that came afterwards, it took ecosystems around 30 million years to fully recover. When they did, the major animal groups were re-arranged, and strange, never before seen reptiles radiated into new ways of life in the sea, on land and in the skies.

Glossopteris – Seed Plants Unite the Planet

The seed plant *Glossopteris* is perhaps the best-known plant in the Permian. Its humble elongated leaves not only tell us about ancient environments, but also provide evidence for the existence of supercontinents and the theory of plate tectonics. Growing throughout the southern hemisphere of the supercontinental world, its fossils are even found on Antarctica – where the first discovery of their petrified leaves were valuable enough to die for.

One plant has come to symbolize the Permian like no other: the tongue-shaped leaves of *Glossopteris*. It is a type of seed-bearing plant, called pteridosperms, that emerged over 375 million years ago in the Devonian and became extinct by the end of the Cretaceous, just 66 million years ago. Pteridosperms were especially common in the Carboniferous and Permian, and *Glossopteris* emerged and thrived in the latter time period. Like so many others, it disappeared in the largest mass extinction of all time at the end of the Permian.

Glossopteris belonged to a woody tree that could reach 30 metres (almost 100 feet) in height. It probably resembled a conifer, but rather than needles it had leaves that ranged from no bigger than your fingertip to the length of your forearm. It is thought that these were shed in autumn. Evidence for a seasonal growth cycle is found in their fossilized wood, which shows signs of increased growth in spring and summer, and dormancy in winter. *Glossopteris* grew in wet conditions, which confined them to the southern hemisphere of the Permian world. Their fossils are found today in Africa, Antarctica, Australia, India, New Zealand and South America, and are a major contributor to coal deposits in these continents.

The name *Glossopteris* only refers to the fossil leaves of the plant. Fossils of its roots are called *Vertebraria*, and the reproductive parts include *Dictyopteridium* and *Ottokaria*. There are many species of *Glossopteris* – as many as seventy in India alone – and researchers are still working on understanding their true taxonomic diversity.

The continent of Gondwana (or Gondwanaland), which had existed since the first complex life began on our planet, was part of the larger supercontinent of Pangaea by the Permian. The idea that the many separate continents on Earth might have once been connected was suggested as long ago as the 1500s, when scientists noted that the shapes of many of their coastlines could be fitted together.

A fossil leaf of *Glossopteris indica*, an extinct type of plant once common across the world's supercontinent.

110

The mechanism for this movement of continents, plate tectonics, was not developed until 1912, by the German polar researcher Alfred Wegener (1880–1930).

A major piece of evidence to support the theory that the continents were once connected came from fossils. It became clear that *Glossopteris* fossils could be found across the southern hemisphere continents despite the fact that they were separated by hundreds of miles of ocean. The seeds of *Glossopteris* were too heavy to be transported by wind, so the most likely explanation was that there had been land bridges connecting these lands in the Permian. This theory was developed by the Austrian geologist Eduard Suess (1831–1914), who coined the name Gondwanaland for the southern supercontinent upon which these fossils thrived and spread.

Glossopteris in Antarctica

Of all the Earth's continents, there is one about which we know comparatively little. Antarctica, which sits on the South Pole and is sheathed almost entirely in a massive permanent icecap, has most of its geology hidden from view. Those rocks that are exposed on its northernmost fringes are difficult to access and scoured by moving ice. Until recently, we knew almost nothing about the history of life in this part of the world.

In 1910 the British Terra Nova expedition, led by Robert Falcon Scott (1868–1912), aimed to be the first to reach the South Pole. Alongside this objective, they intended to collect scientific data, including information on the continent's geology, weather and plants. Although they made it to the pole, they were of course beaten there by Roald Amundsen's Norwegian team, and tragically perished on the return journey.

When the remains of the expedition were discovered eight months later, the fossils of *Glossopteris* were found beside their bodies. Despite abandoning much of their gear to lighten their load as the situation became desperate, the team retained around 15 kilograms (33 pounds) of *Glossopteris* fossils along with other extinct organisms, notebooks and scientific samples. Scott's team had recognized that these specimens and their scientific observations of them were of incredible importance, and chose to keep them even as they struggled for their lives. These fossils revolutionized our knowledge of the frozen continent, proving not only that it was once warm and filled with life, but also that it had been connected to other

Glossopteris-rich continents such as Australia and Africa. As climate change melts Antarctica's glaciers, access to fossils may become easier, though at a far greater and more tragic cost than that of Scott's ill-fated expedition.

Dimetrodon – Ancient Mammal Origins

One of the most misunderstood animals in tetrapod history is surely *Dimetrodon*. This iconic sail-backed creature lived over 270 million years ago in the first half of the Permian. For years it was considered a type of reptile, but we now know *Dimetrodon* and kin were among our ancient relatives. These mammal predecessors were the apex predators of their ecosystems. They gave way to new members of the mammal line as their forest ecosystems collapsed, including the first megaherbivores and the sabre-toothed beasts that preyed on them.

Dimetrodon is instantly recognizable thanks to its striking outline. With a huge sail sticking skywards along its spine like an unfurled fan, it has become a favourite of artists depicting the primeval animals of our planet. *Dimetrodon* is often mistakenly called a dinosaur or reptile, but despite its superficially lizard-like appearance, it is actually one of our early relatives in the mammal lineage, the synapsids. It lived in the first half of the Permian, between 295 to 270 million years ago, and its fossils are found in rocks that were once part of lush river deltas.

Dimetrodon was an ancient relative of mammals. Such creatures were among the most successful tetrapods for millions of years before the dinosaurs existed.

There were multiple species of *Dimetrodon*, ranging from smaller creatures not much bigger than a dog, to monstrous giants longer than a family car. One of their distinct features was a mouthful of pointed, serrated teeth, including enlarged canine-like teeth near the front. This not only tells us that it was a meat-eater, but is also one of the first examples of the more specialized tooth types that evolved in the mammal lineage, allowing them to process different foods more efficiently. *Dimetrodon* probably fed on other vertebrate animals and possibly large insects. Other synapsids ate insects, plants and fish.

Dimetrodon was just one of many synapsids that stalked the Permian. By the middle of this time period, changes in climate dried out the supercontinent of Pangaea, altering the composition of the forests. This coincides with the replacement of animals such as *Dimetrodon* with a new wave of their descendants, called therapsids. These increasingly mammal-like animals continued to multiply in the second half of the Permian, giving rise to the first fast-moving predators and bulky megaherbivores to roam the Earth; a predator–prey relationship that has re-emerged again and again through the millennia, forming the foundation of ecosystems on land.

Not a Reptile

Dimetrodon is often incorrectly depicted living alongside dinosaurs, but these two groups never overlapped; *Dimetrodon* preceded dinosaurs by almost 100 million years. For a long time *Dimetrodon* and kin were known as 'mammal-like reptiles'. They were first recognized as part of the mammal lineage at the end of the 19th century, and it was thought initially that they evolved from a branch of reptiles. However, as more fossils were found and our understanding of evolution and family relationships became clearer, researchers realized that the mammal lineage – called Synapsida – was completely independent from the reptile lineage, with the two branches having evolved from a shared tetrapod ancestor in the Carboniferous.

This misnomer has clung on, despite the fact that we know *Dimetrodon* is neither a dinosaur nor any other kind of reptile. However, it is true that it shares some superficial similarities in appearance. Its limbs were sprawled to the side, and it was hairless and didn't have external ear flaps. It had a long body, and would have walked with a somewhat lizard-like gait (although in truth its movement wasn't quite the same as any reptile). We tend to associate these characteristics with reptiles because they have been lost in almost all mammals, but they are features seen in most early tetrapods, thanks to their shared ancestry.

Sailing Away

The spines that array the back of *Dimetrodon* give it a punk profile, but they also pose a conundrum for scientists. In *Dimetrodon* the spines were thin and tapered like needles, whereas in its close relative *Edaphosaurus*, the spines are covered in knobbly protrusions along their length, like twiggy branches. In large species the spikes could reach over 2 metres (6½ feet) in height. What were these massive structures used for? The first scientists to tackle this question suggested they may have been sails, carrying the animals across the surface of lakes. This was an improbable suggestion, not least because it would require *Dimetrodon* to drift sideways from shore to shore. Other unlikely ideas included that the spines supported fat-storing humps or provided camouflage in dense undergrowth.

The most well-known explanation is that the back sail was used to thermoregulate, assisting the animal in heating up when basking in the sun, and speeding cooling when in the shade. After examining the relationship between body size and sail size, and tracing the patterns of blood vessels in around the spines, there is very little evidence to support this hypothesis. What's more, this explanation relies on the idea that all synapsids were cold blooded. Recent analysis of their bones suggests they may have had faster metabolisms than previously thought, and some may even have been active at night. This means they would not have basked in the sun the way cold-blooded lizards do today.

More recently, scientists have argued that the back sails of these animals played a role in sexual selection and competition, much like the horns of male deer and goats. They may have made the animal appear larger to rivals, or been brightly coloured to attract the opposite sex. Despite more than a century of debate, we still can't be sure why *Dimetrodon* had its long back spines, but they have made it an enduring icon of the pantheon of extinct life on our planet.

Diplocaulus – Mysterious Amphibian Origins

The undulating body of *Diplocaulus* would have been a common sight in the rivers of Permian Earth. While reptiles and the ancestors of mammals took over the wider landscape, amphibian-like animals such as *Diplocaulus* continued to thrive in and around fresh water. Its unique boomerang-shaped skull, perfect for navigating fast-moving streams, made it a swift predator, rising from the riverbed to snatch passing fish and invertebrates. These animals give us clues to the origin of modern amphibians, who now play the same role in modern ecosystems. But tracing their lineage back through time remains a tantalizing mystery – one that science is still trying to solve.

Diplocaulus was a bizarre-headed animal the size of an otter that lived through much of the Permian, from 306 to 255 million years ago. It was amphibian-like, but belonged to a grouping of animals known as lepospondyls, which lived from the Carboniferous into the Permian. *Diplocaulus* had a long tail and wide flattened body, but most unusual of all was its head, which was shaped like a boomerang, with two tiny eyes on the upper surface. Fossils of *Diplocaulus* have been found in what is now North America and Africa. It was the largest of the lepospondyls, at a metre (3 feet) in length, and lived a mostly waterborne, semi-aquatic life, similar to modern salamanders.

The weird skull of *Diplocaulus* gets its shape from two 'horns' that projected backwards from the face. Many theories have been proposed for why it had this V-shaped head. One of the most likely is that the shape was hydrodynamic, allowing the animal to control how it rose and fell in rapidly moving water. This suggests it was a fast-water specialist, able to rise speedily from the riverbed to seize prey from the currents above. Although it hunted in the water, *Diplocaulus* also burrowed in soft muds and aestivated – a kind of dormant period where bodily functions slow down, allowing animals to survive hot and dry spells. The fossils of half-eaten *Diplocaulus* and other amphibian-like animals have been found alongside the shed teeth of the carnivorous synapsid *Dimetrodon*, suggesting it was an important component in the large predator's diet in the Permian.

With so much attention given to ancient mammal relatives and reptiles, it is easy to forget about the many other tetrapods making their home on our ancient Earth. To this day, animals such as salamanders inhabit similar niches in our ecosystems to *Diplocaulus*,

Water-dwelling *Diplocaulus* is one of many animals that provide clues to the origins of amphibians.

118

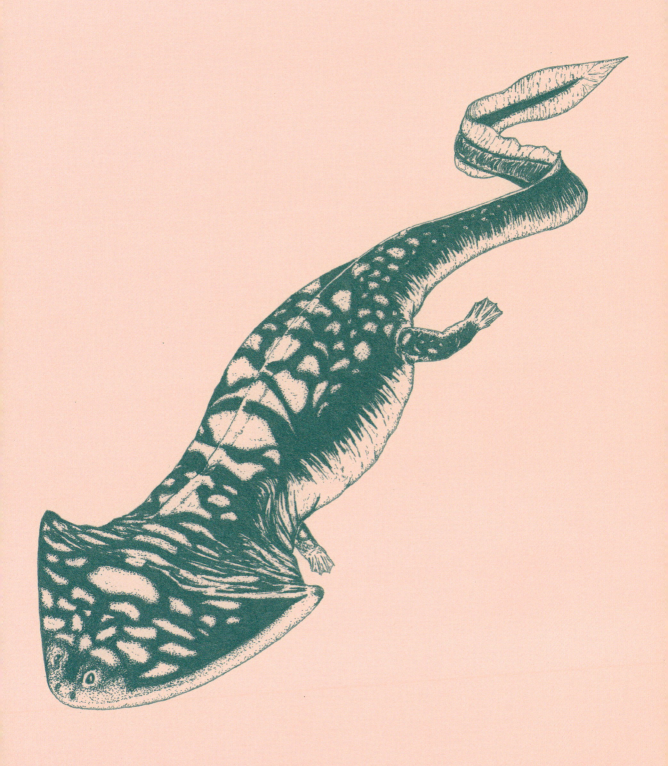

living alongside other tetrapods in rivers and streams, and feeding on fish and invertebrates. Despite over two hundred years of fossil discovery, we still don't know much about the early evolution of amphibians. Studying animals such as *Diplocaulus* helps researchers uncover their story and piece together the emergence of this vital part of life on Earth.

Amphibian Origins

At first glance, *Diplocaulus* looks like a modern giant salamander. These amphibians dwell in rivers in China, Japan and North America, where some grow longer than an adult human. Yet the relationship between *Diplocaulus* and today's living amphibian groups is somewhat puzzling.

There are as many as eight thousand species of amphibians today, the group that includes frogs, salamanders and caecilians (a worm-like burrowing animal). We know from DNA that they share a common ancestor as far back as the Carboniferous, but their first fossils are known only from the Triassic. Lepospondyls such as *Diplocaulus* lived alongside another group called the temnospondyls, and although referred to as amphibians, debate continues as to which – if either – are actually the ancestors of those alive today.

Ecological Indicators

Amphibians have traits that make them unique among living tetrapods. They are anamniotes, meaning that they produce simple eggs without a tough outer shell, and usually rely on fresh water for reproduction (they aren't usually found in salty environments, although there are one or two exceptions). Their young go through a radical metamorphosis, from water-dwelling larvae to air-breathing adults. Some forgo part of this process, remaining water-dwellers and retaining their gills, such as the axolotl (*Ambystoma*). Most develop lungs and also respire through their mucous-rich skin – there are even species that lack lungs entirely and rely solely on their skin to breathe. Some are poisonous. Although often associated with ponds and streams, amphibians have adapted to water-poor conditions, and others can leap great distances and climb trees, even moving between distant branches using their webbed feet to glide.

The unique physiology of amphibians makes them especially sensitive to environmental changes. For this reason they are considered to be 'ecological indicators'. This means that their presence provides a short-hand for the health of an ecosystem or habitat.

Their semi-permeable skin and need for fresh water for reproduction and growth make them susceptible to pollution and habitat loss, as well as changes in the food web caused by removal or introduction of other species. As they are often prey for larger animals, their disappearance can have a radical effect on the healthy functioning of ecosystems.

Deadly Fungus

The most recent and devastating impact on amphibian numbers comes in the form of an infectious fungus called *Batrachochytrium*. Also known as chytrid fungus, this has affected over a third of amphibian populations worldwide. It causes thickening of the skin, affecting the animal's ability to breathe, and often makes them lethargic and slow, impacting their ability to escape from predators.

Although the origin of the fungus is not clear, it is being spread more widely by international trade in amphibians as pets, for aquariums and for research. Climate change is thought to be speeding the spread. In some areas, chytrid fungus has had up to 100 per cent mortality rate, decimating a key component of the ecosystem. Researchers have calculated that current extinction rates among amphibians are accelerating, and may be as much as 45,000 times faster than normal. Despite surviving many natural mass extinctions through geological time, it is likely that humans pose the greatest threat this ancient lineage has ever faced.

Conifers – The Toughest Trees

Conifer trees make up the majority of woodland on our planet today. It was in the Permian that they began to diversify, taking advantage of the drier climate of the newly melded supercontinent. In the aftermath of the end-Permian mass extinction, they bounced back rapidly to become the most important tree on Earth. From dinosaur food, to mythical symbol of immortality, to commercial lumber crops, the conifer is integral to the story of life on Earth.

The largest single biome on our planet today is the taiga, or boreal forest. It coats most of the northern hemisphere in a dark green pelt, from Russia and northern Europe to North America. These forests are dominated by conifers, cone-bearing plants that make up the majority of trees on Earth. Although an integral part of our planet now, the ancestors of conifers appeared only around 300 million years ago in the Late Carboniferous. It was in the Permian that they truly began to flourish, going on to become the main food source for many herbivorous dinosaurs in the Mesozoic. Fossils from the time of their first emergence are sparse, comprising mainly fragments and pollen.

Conifers are mostly trees, although a few are shrubs. They include the tallest trees in the world (*Sequoia*), and generally have long needles, flattened scales or strap-shaped leaves. Cypress, fir, juniper, kauri, larch, spruce, pine, redwood and yew are all conifers, and they are found on almost every continent. Conifers have one of the largest genomes of all organisms on the planet. With over 650 species, they are not the most diverse type of plant, but they cover massive tracts of land. They are vital to life on Earth, including humans. They constitute one of the largest carbon sinks, making them important in our battle against human-induced climate change. They are of huge economic value for timber and paper production, providing almost half of the world's annual lumber. They are key for other products such as soap, food, perfume, nail varnish and gum. Not only this, but their branches are also woven into human heritage and cultures around the world, from their use as longbows as much as five thousand years ago, to their continued association with winter festivals, as symbols of endurance through hard times.

Conifers are gymnosperms, meaning 'naked seed' in Greek. This group includes cycads, ginkgoes, gnetophytes and pines, all of which have seeds that develop on the surface of scales, leaves or in cones (as

Conifers come in many forms, and include the oldest living trees on Earth.

opposed to being enclosed within an ovary, as in angiosperms, the flowering plants). There are many extinct types of gymnosperm, such as the bennettitales, which were prolific in the time of dinosaurs. Conifers are by far the most common gymnosperms on Earth today, but others, such as the ginkgoes, are sole surviving species and face an uncertain future.

Surviving Extremes

By developing pollen that could be carried on the wind, and seeds protected by cones, conifers were able to withstand much drier climates than the types of trees that had comprised the majority of the first forests on Earth. As climates changed in the later part of the Palaeozoic, conifers and other seed-producing plants (gymnosperms) had the advantage, and spread throughout the landscape. At the end of the Permian, the largest mass extinction of all time dealt a bruising blow to life on Earth, but conifers quickly replenished in the aftermath. The earliest pine fossils come from the Late Triassic, after which conifers are abundant in the fossil record. Although conifers declined at the end of the Cretaceous, as new types of plant and tree evolved, they nevertheless remain an integral part of our planet's ecosystems.

Conifers are gloriously hardy to extreme conditions. They are often associated with high latitudes and altitudes. Many have specialist adaptations for colder environments, including drooping branches to shed snow. In high latitudes they are usually darker green than other trees because they are packed with photosynthesizing chlorophyll to squeeze out the maximum energy from weaker, more infrequent sunshine. Conversely, in sunnier climes conifers often have a silvery sheen, protecting them from ultraviolet rays. All of this makes them true survivors.

Oh Tannenbaum

One of the most evocative traits of the conifer is its heavily scented resin. This substance is usually secreted when a tree is injured, and helps protect them from insects and fungal infestations. Some resin scents even attract other invertebrates that will eat the plant's attackers. As well as being useful in human medicine and perfumes, resin can fossilize to form amber, which has been used for jewellery for at least 13,000 years.

The fact conifers remain green in winter when all other life dies has made them a symbol of endurance and eternal life around the world. This connection is warranted: the ten oldest trees alive on our planet today are all conifers, the oldest being a great bristlecone pine (*Pinus longaeva*) in Nevada, United States, at over 4,900 years in age.

MESOZOIC

The Mesozoic covers what are perhaps the three most famous time periods in Earth's geological history: the Triassic, Jurassic and Cretaceous. It lasted from 252 to 66 million years ago, beginning and ending with mass extinctions that reversed the roster of our planet's dominant animals. The Mesozoic goes by many monikers, including the 'age of reptiles', because for 186 million years they flourished in the skies, seas and on land, growing to become the largest animals ever to walk the Earth. Despite appearing unimaginably strange to us, the Mesozoic heralds the 'birth of modern ecosystems'. The planet itself took on recognizable continental outlines, and for the first time we see members of all the major groups of organism that continue to share our world.

The Mesozoic literally grew from the ashes. At its dawn, a series of massive volcanic eruptions decimated the land and poisoned the skies and seas, resulting in a global mass extinction that killed over three-quarters of all life. In the oceans, marine life faced acidification and lack of oxygen. Charismatic groups such as trilobites and eurypterids (sea scorpions) were lost forever. On land, the synapsids (mammal line) that had flourished since the emergence of four-limbed animals from the water's edge were reduced to only a few lineages. From pole to pole, acid rain and nuclear winter brought life to its knees.

This wholesale destruction created new opportunities. A few hardy generalists took advantage of their vacant world, and ecosystems slowly recovered. A different branch of tetrapods came to dominate the planet: the reptiles. The 'time of the dinosaurs' began. They included giant behemoth herbivores that dwarfed all life before and since, meat-eaters with teeth like carving knives, and small feathery flappers – the ancestors of birds. Their fossil bones have long been incorporated into folklore in Asia and the Americas, before amazing European scientists in the 18th and 19th centuries. Although reptiles steal the Mesozoic show, other world-changing lineages were shaped at this time. The common ancestors of modern mammals emerged, along with crocodiles and amphibians. Among insects, beetles suddenly diversified, and the ancestors of butterflies, ants and bees appeared during the Cretaceous Terrestrial Revolution. This heralded a development that reshaped life on Earth: the appearance of the first flowering plants.

Meanwhile, the supercontinent of Pangaea shattered like a dropped dinner plate, forming the beginnings of our modern geography. Sea levels rose and fell, creating new shorelines and altering climates. In the warmth of the Cretaceous waters, marine plankton rained down on the shallow seabed, generating iceberg-like chalk beds hundreds of metres deep. At the end of the Mesozoic, the world was pummelled by an asteroid impact that unbalanced the web of life once more. The killer impact turned the page on animal life – ushering in a new 'age of mammals'.

Triassic

252 to 201 million years ago. The Mesozoic Marine Revolution began and multiple reptiles returned from land to life in the sea.

Sea levels were generally low.

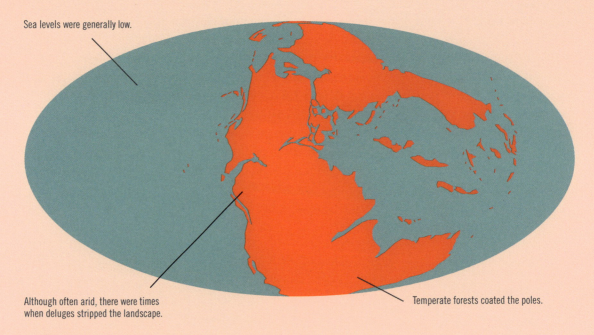

Although often arid, there were times when deluges stripped the landscape.

Temperate forests coated the poles.

Jurassic

201 to 145 million years ago. Dinosaurs proliferated on land, as did mammals, small reptiles and insects.

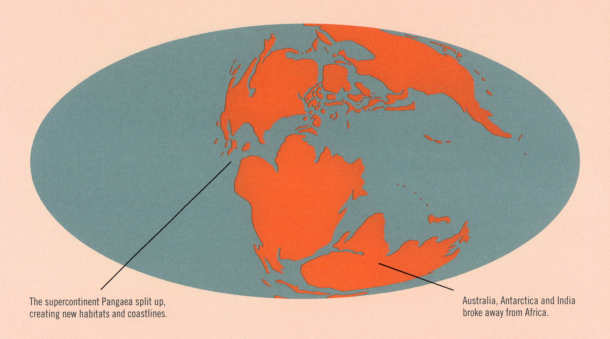

The supercontinent Pangaea split up, creating new habitats and coastlines.

Australia, Antarctica and India broke away from Africa.

Early Cretaceous

145 to 100 million years ago. The Cretaceous Terrestrial Revolution began on land, triggered by the emergence of flowering plants.

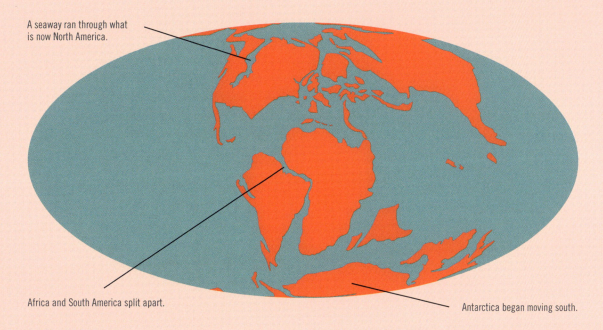

A seaway ran through what is now North America.

Africa and South America split apart.

Antarctica began moving south.

Late Cretaceous

100 to 66 million years ago. Sea levels were over 110 metres higher than present.

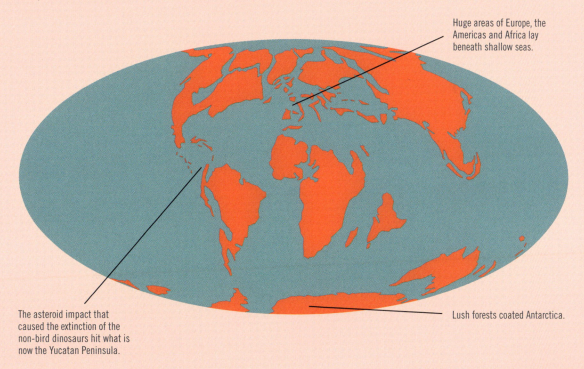

Huge areas of Europe, the Americas and Africa lay beneath shallow seas.

The asteroid impact that caused the extinction of the non-bird dinosaurs hit what is now the Yucatan Peninsula.

Lush forests coated Antarctica.

Triassic

The brief and wacky Triassic swirled into being from the dust of the largest mass extinction of all time. It was a period of recovery and invention. As animals and plants diversified again, they explored new ways of life. On land, sea and in the air, reptiles took leading roles in the forms of dinosaurs, marine reptiles and pterosaurs. Beneath their feet, the Earth itself was re-engineered, from the soil to the tectonic plates themselves.

The Triassic began 252 million years ago, lasting just 51 million years. It was a short but wildly inventive time, when reptiles took centre-stage in the evolutionary story. The supercontinent of Pangaea remained, and there were baking hot summers and cold winters inland, while monsoons inundated the coasts. The rocks from this time period tell of episodes when rainfall increased so much that huge volumes of soil and rock were washed away. Although sea levels were generally low, during the times of excessive rainfall they rose as much as 50 metres (165 feet) above today's level. Temperate forests clothed the poles, including conifers and bennettitales: a group of plants that resembled palms, with long, thin leaves. Now extinct, they were the most common plants throughout the Mesozoic.

The colourful tapestry of diversity that had been woven throughout the preceding Palaeozoic was burned to dust by the start of the Triassic. Volcanoes in what is now Siberia erupted millions of tonnes of molten rock and poisonous gases, smothering an area the size of Australia in lava. More destructive than the lava were the gasses vented into the atmosphere. Carbon dioxide made a stultifying greenhouse of the planet. Sulphur bound with moisture droplets to create burning acid rain, killing plants and acidifying rivers and seas. Between 70 to 80 per cent of animals on the planet became extinct. It took as much as twenty million years for life to recover.

The patterns of Triassic life's recovery help us understand the effects of our current extinction crisis. Not every creature disappeared immediately – some held on through the worst of it but petered to extinction after a few million years. They illustrate the lag between an extinction-causing event and the final disappearance of organisms themselves. Some creatures flourished to begin with but disappeared in the longer term – so-called 'disaster taxa'. One of the most famous is the pig-like herbivore called *Lystrosaurus*, an ancient relative of mammals. This is one of the only times in the history of life where a single group dominated every continent.

As the world came back to life, forests of conifers, ferns and palm-like bennettitales returned. Novel organisms occupied the empty niches, and animals such as *Lystrosaurus* were supplanted. The roster of new animal groups is staggering: the first marine reptiles, dinosaurs, mammals and modern fishes all appeared. Pterosaurs evolved, the first flying tetrapods in our planet's history. It is often considered a time of mind-boggling experimentation for organisms, and these new creatures made the Mesozoic their own.

Marine Revolution

In the Triassic a revolution also took place beneath the surf. The appearance of new groups of organisms in the oceans was part of an event known as the Mesozoic Marine Revolution. New types of coral provided shelter for the first modern lineages of fishes, which now comprise the greatest vertebrate diversity on the planet. But it was tetrapods that unexpectedly capitalized on the wealth of the Triassic oceans. Four-limbed animals returned to the sea; first as visitors but eventually as permanent residents. These air-breathers underwent whole-body modifications to adapt to marine living: losing their limbs, forming paddles and giving birth to live young.

Although animals had been feasting on one another since life began, after the Triassic there was an increase in shell-crushers. Invertebrates that once seemed safe in the vault of their shells became prey to marine reptiles that evolved species with hard, flattened teeth and strong jaws to break open shellfish. Animals that were fixed to the seabed or moved very little were easy pickings, including crinoids (relatives of sea urchins) and star fish. In response, some shelled animals evolved armour defences such as spikes or became increasingly mobile to escape the myriad biting jaws.

End-Triassic Turnover

There were in fact many extinctions and new beginnings in the closing part of the Triassic. An event called the Carnian Pluvial Episode left its mark in the rock record with huge volumes of sediment, carried by water running off the continents and filling valleys and deltas with debris. This change in climate was sudden and reversed just as quickly, but it has been tied to the evolution of multiple animal groups, including dinosaurs and smaller reptiles such as the ancestors of lizards and snakes.

In the first fifty million years of the Triassic, the ancestors of crocodiles did better than dinosaurs. These crocodiles grew longer than a car, walking tall on long, powerful sprinting legs, and some were even bipedal. Yet their reign was not to last: another round of volcanic eruptions cleared a path for the dinosaurs to complete their takeover and prosper for the next 150 million years.

Meanwhile, the precursors of mammals discovered an altogether different niche. They evolved fur and milk production. Their adaptations for being small nocturnal insect-hunters drove anatomical changes that set the groundwork for their incredible diversity later in Earth's history. At the end of the Triassic, they too experienced a new lease of life, accompanying the giant reptiles into this brand new future.

Pholidophorus — Origin of Teleost Fish

Fish are one of the most important food sources in the world. Although they first emerged in the Devonian, it was in the Triassic that most of the fish that shoal along our shores, leap in our rivers and lurk in our lakes today owe their origins. *Pholidophorus* was one of the earliest modern fish, with a new jaw arrangement that gave it the ability to bite swiftly using suction. It was this that helped make fish the most diverse vertebrate group on Earth today.

Swimming in the warm Triassic seas was a piscene resident of staggering import. *Pholidophorus* looked like a herring and grew to a similar size – not much bigger than your forearm. It had a long body and shimmering scales, and its tail base was narrow, flaring out to form the pointed halves of a symmetrical fin. This ancient fish, common over 200 million years ago, may not seem much different

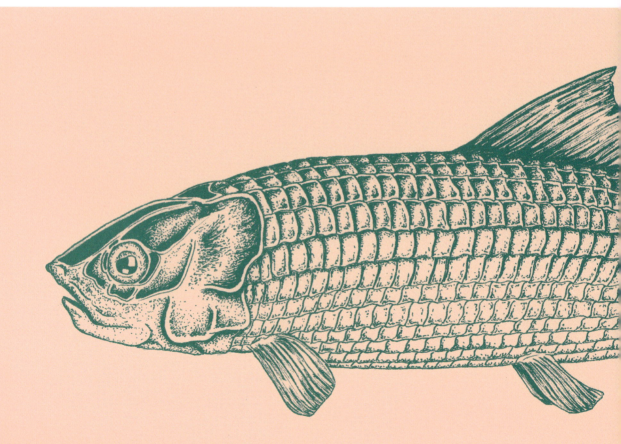

from those that came before, but unlike earlier scaly denizens of Earth's waters, *Pholidophorus* belonged to a vibrant new group that would become the most successful and speciose vertebrate group on our planet: the teleost fish.

Pholidophorus has features that reveal its position at the very base of the teleost fish family. This makes it the relative of almost every living fish on Earth, and crucial for understanding the early history and diversification of modern groups. Its beautiful scales slid through the seas off the coast of what are now Africa, Europe and South America, and they continue to glitter in the fossil record. Many specimens are beautifully preserved, looking as if they might flop off and swim away at any moment.

Teleost fish make up 96 per cent of all living fish species, and are found in almost every aquatic environment, even in extreme conditions like hot, ultra-saline desert lakes and cold, isolated cave pools.

Pholidophorus was one of the first teleost fishes, the modern group to which most of today's fish belong.

Some are migratory and travel incredible distances; such as the eels (*Anguilla*) that undulate over 3,700 miles (6,000 kilometres) across the Atlantic Ocean to spawn. They are an important part of aquatic ecosystems the world over, as well as providing a food source to over three billion people every year. Some, such as the zebra fish (*Danio rerio*), are also crucial to medical science. Not only are zebra fish considered model organisms for understanding vertebrates – being used in studies of animal development, common medical disorders, gene expression and in creating and testing new pharmaceuticals – but also, in the 1970s this humble species was one of the first backboned animals to be cloned.

Jutting Jaws

There are over 26,000 species of teleost fish today, ranging from the size of a paperclip, to the gargantuan sunfish (*Mola mola*), which weighs more than 15 adult men, making it the heaviest bony fish on the planet. What distinguishes a teleost fish from other kinds lies in the architecture of the jaws. Teleosts can protrude the front part of their jaws forwards, grabbing their food and bringing it back to the mouth like a hand greedily darting out at a buffet table. The sudden jutting action also produces suction, pulling the prey towards them. This transformed the speed they could obtain food, making them formidable feeders. As the Mesozoic Marine Revolution picked up pace, this new group was ready to exploit every new opportunity.

Most teleost fish have colour vision, and many can sense chemical 'tastes' in the surrounding water, or movement and vibration through an organ called the lateral line, which runs the length of the body. They are cold-blooded, but there are species with higher metabolisms, such as swordfish and tuna. Teleost fish have an extraordinary number of reproductive strategies. Their sex is sometimes determined by their environment, and in species such as the clownfish, individuals can change sex, for example when the dominant male in a breeding group dies. Although most lay eggs and fertilize them externally, there are those that retain their eggs, and some species, such as freshwater arowanas, protect young by holding them inside their mouths. When the small fry are ready to leave, that jutting jaw works in reverse, ejecting them into the water to make their way in the dangerous depths.

Go Fish!

As they dominate the oceans and waterways of Earth, teleost fish comprise all the major human fisheries catches globally. Over 96 million tonnes are caught or farmed each year, providing the main source of protein for over three billion people. Despite their amazing variety, in the last few centuries numbers have collapsed dramatically. Overfishing, coupled with discarding bycatch, has steadily depleted fish numbers. Enormous industrial fishing vessels have impacted shoals that once fed hundreds of small-scale fishing villages and towns around the world. Trawlers rip up the seabed, hindering the recovery of marine life by removing spawning and nursery grounds. Climate change and pollution have also taken their toll, particularly on coastlines and inland waters. Action is being taken to combat the downward trend: marine reserves and changes in fishing practice are among the network of solutions being adopted to combat the problem. Although in some places there are small signs of recovery, stocks are still far below their historic bounty. The overfishing crisis has been identified as one of the greatest threats we face today.

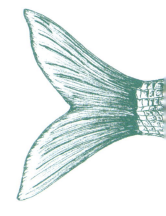

Utatsusaurus – Returning to the Sea

In the Early Triassic ocean over what is now Japan swam *Utatsusaurus*, a creature the size of a dolphin that fed on fish. It belonged to one of many reptile groups to take a startling evolutionary U-turn: returning from their life on land to become fully marine. The pressures of life beneath the waves transformed them inside and out, shrinking limbs and flattening tails. The discovery of their fossils changed human understanding of the past.

In the Triassic, Japan lay under a tepid sea. Swimming among the ammonites and fish shoals were creatures that looked similar to dolphins but were the first of an epically successful group of marine reptiles. *Utatsusaurus*, named for the town of Utatsu-cho, was an ancestor of the ichthyosaurs. It was around 3 metres (10 feet) from nose to tail tip. Its body was smooth with four stout flippers, and it had a long head with an elongated fish-catching snout. Unlike later ichthyosaurs, *Utatsusaurus* had an eel-like tail. Ichthyosaurs were one of the many marine reptiles that made the Mesozoic oceans their home.

Marine reptiles are among the first backboned animal fossils found in the Mesozoic rocks of Europe. Mary Anning (1799–1847) made many of the earliest discoveries on the beaches of Lyme Regis in England. So many marine reptiles were found, naturalists assumed the world was mainly covered in water during the 'age of reptiles'. As understanding improved, it was soon realized that the abundance of marine animals had more to do with the rocks people were looking in – which were formed at the bottom of the sea – than the makeup of the planet's surface in the past.

Although often mistakenly lumped in with the dinosaurs, marine reptiles belong to completely independent lineages that share common ancestors as far back as the Permian. Some of the other groups that adapted to life in the Mesozoic seas include: plesiosaurs, with their four flippers, long necks and tiny heads; powerfully built mosasaurs; marine crocodile relatives called thalattosuchians; and bizarre duck-billed hupesuchians. Their origins are often a mystery because their ancestors adapted so rapidly that they left very few fossils in their wake. Being reptiles, it was assumed that these creatures laid eggs, perhaps dragging their hefty bodies onto the beach to deposit a clutch, as turtles do today. We now know they gave birth to live young in the water – as dolphins do – severing all ties to the

This old engraving of *Plesiosaurus* (behind) and *Ichthyosaurus* (front) from the 19th Century reflects our longstanding fascination with ancient marine reptiles.

land their ancestors once stalked. Several fossil ichthyosaurs are preserved with embryos inside the womb or in the birth canal. Unlike whales and dolphins, ichthyosaurs gave birth to large litters; a fossil of *Stenopterygius* has 11 foetuses preserved inside it.

Return to the Sea

The astonishing return to life in water from land completely reshaped animal bodies. It has occurred multiple times in different groups, including reptiles, and mammals over 200 million years later. The very first reptiles to return to water did so in the Permian, but in the Triassic more followed suit. They evolved from small lizard-like creatures that stayed close to shore, into ocean-going monsters up to 20 metres (over 65 feet) in length – only slightly shorter than a blue whale. Many of the changes we see in marine reptiles are charted in the fossil record, and later echoed in the evolutionary story of whales and dolphins. Their limbs shortened, steadily selected to become paddle-like, or were lost altogether. Their bodies streamlined to reduce drag and increase swimming efficiency. In groups such as ichthyosaurs, their tails flattened and were used to power propulsion.

There were internal changes, too. In some marine mammals today, a single breath can last up to two hours because oxygen is not only packed into their lungs, but is also stored in muscles and other tissues, helping to fuel deeper, longer dives. This may have also been true for marine reptiles, which are thought to have been warm-blooded (endothermic) like their cousins the dinosaurs and pterosaurs. Deep-diving ichthyosaurs had the largest eyes compared to their body mass in the history of life on Earth; their peepers were as big as dinner plates, perfect for hunting in the darkest depths. Decompression sickness was a real danger for these divers, as it is for modern whales. This condition is caused when gas bubbles form in the body if it rises to the surface too quickly. Fossils of marine reptiles show evidence of decompression sickness, which leaves pitting in the bones as cells are damaged and die.

Loch Lurkers

Not only do marine reptiles survive as fossils, they also live on in our myths and imaginations. Legends of lake and sea monsters have been attributed to lone survivors from the time of dinosaurs. These tales often predate common knowledge of marine reptiles, and the animals were originally considered to be monsters or mythical beasts. As fossil discoveries have increased, they have been reclassified as types of Mesozoic marine reptile – often long-necked, small-headed plesiosaurs. Examples include the Ogopogo of Okanagan Lake in Canada, and the Phaya Naga in Southeast Asia. Perhaps the most famous is the Loch Ness Monster. Loch means 'lake' in Gaelic, the indigenous language of Scotland and Ireland. First mentioned in the 6th century, it became popular in the 1930s when photographs surfaced of a long-necked animal lurking in the water. Despite countless searches, no concrete evidence has ever emerged to prove that there is a strange animal living there.

Although it is alluring to imagine these creatures are the descendants of Mesozoic marine reptiles, there are many practical reasons why this is unlikely. There is no fossil evidence that any marine reptiles survived the mass extinction 66 million years ago. It would have taken millions of generations to sustain a species into the present day, so their existence would be difficult to overlook. The lakes these legends inhabit are geologically recent features; for example, Loch Ness was gouged out during the last ice age, which ended only 10,500 years ago. Despite the science to the contrary, stories of these creatures are unlikely to ever become extinct.

Postosuchus — Origin of Crocodiles

Postosuchus was a distant crocodile relative bigger than a tiger, which prowled our Triassic planet for over twenty million years. It belonged to a plethora of animals that initially outcompeted the dinosaurs, including armoured, tank-like beasts and sprinting hunters. Their lineage would go on to produce the ancestors of modern crocodiles, alligators and gharials, the only survivors from a once-bushy family tree.

Postosuchus was an ancient reptile that prowled the North American continent 222 to 202 million years ago. It grew as long as a car, with an enormous skull full of serrated teeth. The strong hind legs and smaller forelimbs suggest it may have even walked on two legs. Although at first glance this fearsome animal resembles a predatory

dinosaur, it was more closely related to modern crocodiles, belonging to a group called the rauisuchidans. They included the largest predators on land in the Triassic, and their fossils can be found throughout the northern hemisphere, South America and South Africa.

Fossils of *Postosuchus* are found in what was once a warm tropical environment, lush with ferns. Although *Postosuchus* was large, it was by no means the biggest rauisuchidan in the Triassic. Some grew to 6 metres (20 feet) from tooth to tail, and all stood on strong, upright legs – not sprawled to the side like today's crocodiles. These were fast-moving predators, feeding on other vertebrate animals. *Postosuchus* belonged to the most successful lineage of reptiles on land for most of the Triassic. Only at the end of this time period did these crocodile relatives give way to the rising ranks of dinosaurs.

The ancient relatives of crocodiles included sprinting predators with long limbs. Their skeletons and even fossil footprints are known from Triassic rocks.

Although less diverse now than in their Mesozoic heyday, crocodiles (or more correctly, crocodilians) are undoubtedly one of nature's great success stories. Having survived multiple mass extinctions, they continue to thrive in today's watercourses throughout the tropics in the form of crocodiles, alligators, caimans and gharials. These modern groups share a common ancestor around 120 million years ago. They now inhabit mostly freshwater environments, but can also live in brackish and marine waters. They range in size from the dwarf caiman (*Paleosuchus palpebrosus*), at just over a metre (3 feet) in length, to the behemoth saltwater crocodile (*Crocodylus porosus*), which reaches in excess of 8 metres (26 feet). Although cold-blooded (ectothermic), crocodiles can move exceptionally quickly. All are predators, often waiting motionless for hours, days or, in some cases, even months for their next meal to stray near enough to grab.

The Ruling Reptiles

The two major reptile groups that emerged in the Triassic can be split into the 'crocodile line' (including *Postosuchus*), and the 'bird line' (including dinosaurs). Animals such as *Postosuchus* were not technically crocodiles, but they share a common ancestor with them. This group was staggeringly diverse in the Mesozoic: there were massive apex predators and scavengers, as well as small, darting species no bigger than a greyhound. Some were bipedal, while others remained on all fours. There were also plant-eaters that reinforced their bodies with armour and enormous defensive spikes longer than a baseball bat. The bird line, on the other hand, includes all animals more closely related to birds than to crocodiles. This encompasses the dinosaurs, as well as the first flying vertebrates, the pterosaurs.

Altogether, the crocodile line and the bird line are known as archosaurs, which means 'ruling reptiles'. Understanding the evolutionary relationships between archosaurs is a huge part of palaeontology. They not only comprised the most prolific and exciting animals in the Mesozoic, but also continue to flourish throughout the world today.

Triassic Turnover

At the end of the Triassic, one of the five largest mass extinctions of all time devastated the ancient crocodile relatives. It is thought to have been caused by volcanic eruptions in the centre of the supercontinent Pangaea. Like Siberia at the end of the Permian, this part of the world was smothered by hundreds of miles of flood lavas, covering parts of what is now northern Africa, Brazil and western Europe. These eruptions changed the climate, triggering the disappearance of species around the world as food webs were disrupted.

In the havoc, all of the crocodile line were wiped out, except one branch. With their main competition removed, the dinosaurs quickly proliferated, populating every continent in the Jurassic with creatures of unimaginable shape and size. The crocodile line still did well in the Jurassic and Cretaceous, evolving semi-aquatic creatures similar to those alive today, as well as marine species. Only much later were these groups winnowed to the small number of sprawling animals that now lurk in our rivers and watering holes, dragging unsuspecting animals to their demise.

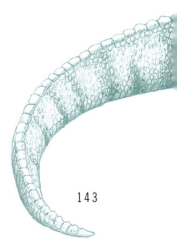

143

Earthworms — Soil Engineers

Earthworms are the subterranean engineers of our planet. Their long history is hard to trace because their soft bodies rarely fossilize, but the field of genetics is revealing the secrets of their evolution. As Pangaea broke apart in the Late Triassic, earthworms were split into two major groups that have worked the fertile earth of their homelands ever since. Recycling soil nutrients, they create the base for almost all ecosystems, and modern agriculture as we know it.

The humble earthworm, such as this *Lumbricus*, engineer entire ecosystems as they process the soil.

There may be no animal on Earth as underappreciated as the glorious earthworm. They are a type of annelid, the group that includes ragworms and leeches. Worms have a poor fossil record because their soft bodies are much less likely to be preserved than animals with hard skeletons. The very oldest annelid fossils are around 520 million years old, but there are currently no known earthworm fossils. Figuring out when, how and where these soil-living species evolved has proven extremely tricky, but we now know it was in the Latest Triassic that the humble earthworm first got to work. There are over six thousand species today, and their activity makes them crucial terrestrial ecosystem engineers; aerating soil, dragging plant matter from the surface deep into their burrows, and making nutrients available to plant and animal life. Their appearance transformed soil fertility, and it is likely that without them, our world would be an impoverished echo of the one we know today.

Being small legless soil dwellers, it takes a long time for earthworms to reach new habitats. Their presence in nearly every part of the world suggests they have a very long evolutionary history. By analysing earthworm DNA, scientists have discovered that modern species form two main groups: those common in the southern hemisphere, and those in the north. They've traced this split to the breakup of the supercontinent of Pangaea, which began in the Triassic. This means that modern earthworms evolved at around the same time as the first dinosaurs.

These incredible animals are more than just a squishy tube. Earthworm bodies are composed of segments, each with its own nerve system interconnected to the whole. Their circulatory and digestive system runs the entire body length, and they breathe through their moist, sensitive skin. The salt in even the gentlest human touch is toxic to them. They have no eyes, but cells in their skin detect light. Earthworms are hermaphrodites, possessing

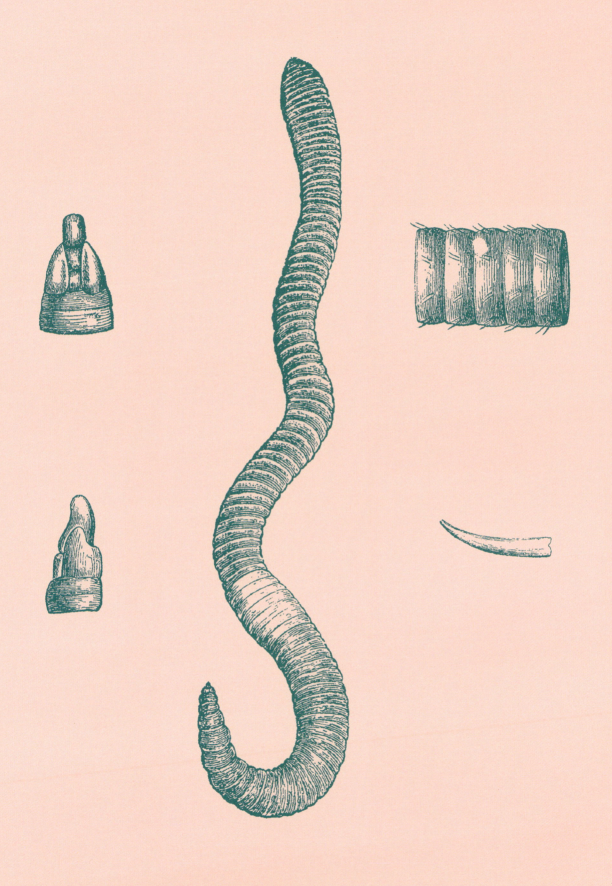

both male and female sex organs. Although most species could comfortably coil in the palm of your hand, they range from the titchiest 1-centimetre miniatures, to giants up to 3 metres (10 feet) long. Earthworms are prey for multiple species of animals, including humans. In Māori culture some are considered a delicacy, called *noke*. Once the food of chiefs, their consumption as a fashionable wild food is now spreading. Humble it may appear, but the earthworm not only has an ancient pedigree, but also a core place on our biodiverse planet.

Dispersal or Vicariance

The study of where organisms live and how they got there is called biogeography. Organisms can spread from one place to another by walking, swimming or crawling (called dispersal), or populations can be physically split or pulled apart, for example by plate tectonic movements (called vicariance). The latter is how earthworms have ended up on every continent. Many animal family trees can be explained by changing geography. When a river forms or a mountain range thrusts upwards, the populations on either side of the divide no longer interbreed. Natural selection works differently on each group, leading to new animals often quite unlike their common ancestors.

Despite seemingly insurmountable odds, another way animals can reach new continents is by 'rafting'. This is when large chunks of soil are dislodged – perhaps by floods or severe storms – and drift out to sea. Ocean currents carry them to new shores, like life rafts, bringing any inhabitants that survive the journey with them. This is how some animals have reached isolated islands. Rafting has been suggested to explain the biogeography of another soil-dwelling group, the amphisbaenians, a type of legless lizard that resembles a snake. Some animals can also be carried to new lands on the feet of birds, particularly in the form of their unhatched cocoons.

Ecosystem Engineers

Earthworms rework soil, like tiny ploughs. This process of burrowing through layers of the earth is called bioturbation. Although earthworms are not the only animals that do this (all animals that dig, from shrimps to rabbits, are bioturbators), their global distribution makes them perhaps the most important. At times in Earth's history, new types of bioturbators have evolved both in the sea and on land, often completely transforming ecosystems by remodeling food webs from the ground up.

Charles Darwin praised worms, recognizing their essential role in creating healthy soils. They are known as soil engineers because their presence affects the properties of the earth itself. They mix layers and make passages for air and water to circulate. They bury archaeological remains, and bring them to the surface again. Worms are vital for human agriculture around the world. In South America, worms in wetland and forested areas create mounds that rise above the water level, creating new habitats. When indigenous peoples built similar raised beds for planting, the earthworms moved into them, breaking up the soil and making it even more fertile.

Despite their incredible importance, earthworms are hugely underappreciated by humans. Many soil fertilizers added to improve harvests kill earthworms, reducing the long-term health of the very ground they are supposed to be improving. This practice has nearly wiped out the rare Karmai, or Gippsland earthworm (*Megascolides australis*), a metre-long (3 feet-long) giant, native to Australia. One estimate suggests that thanks to the earthworm's constant work making the earth fertile for crops and livestock grazing, every person on the planet is supported in kind by the work of around seven million earthworms.

Jurassic

The Jurassic is a byword for the prehistoric, but despite being synonymous with dinosaurs, this period was an incredible time in the history of multiple animal groups. It lasted 56 million years, picking up after the end-Triassic mass extinction. As the supercontinent cracked apart, it created ecosystems that – despite the giant reptilian inhabitants – we might finally call 'modern'.

The Jurassic began 201 million years ago and lasted until the Cretaceous began. The two main continents of Laurasia and Gondwana continued to fragment throughout the period. Global sea levels began to rise, and newly forming oceans produced hundreds of miles of shallow marine and coastal habitat, and altered the global climate. The Jurassic world was warm on land and in the sea, hot-housed by higher carbon dioxide levels than we know today. In the seas there was a regime shift in ocean chemistry, as the first calcareous plankton emerged. These tiny floating organisms, by using carbon from the atmosphere and ocean to build their bodies, stabilized the Earth's biochemistry and reduced the impact of environmental changes on the marine world.

The Jurassic was an incredible time for reams of animal families, giving rise to groups that persist today. The first true crabs evolved, becoming a staple of our seas and freshwaters. Marine reptiles reached the peak of their diversity by taking advantage of similar explosions in species of fish, and marine invertebrates such as belemnites and ammonites. On land, conifers became the dominant forest-forming plant around the world. Their boughs hung over an understorey of ferns and palm-like bennettitales. Among the foliage, new regimes of dinosaur flourished. With the crocodile-line reptiles knocked from their top spot, there was space for dinosaurs to evolve into an incredible plethora of sizes and shapes. The three main branches of their massive family were founded: the long-necked sauropods, the smaller plant-eating ornithischians and the bipedal theropods. From the latter group, the first birds took flight, establishing themselves as a core component of modern ecosystems.

Skies Come Alive

Until the start of the Jurassic, the only creatures to evolve powered flight were insects. They had swarmed the skies since the Carboniferous, their only airborne predators being their larger insect kin, such as dragonflies. At the end of the Triassic, however, vertebrates pursued them skywards. Pterosaurs were flying reptiles related to dinosaurs. They first appeared in the Late Triassic, but it was in the Jurassic that they evolved from small, blunt-headed animals flapping among the trees into species living in many habitats and feeding on everything from small insects and fish, to other reptiles and mammals.

Later in the Jurassic, another group of reptiles joined pterosaurs in flight. Some small, feathered dinosaurs expanded the long feathers on their forelimbs to create a wing. The abundance of flying insect food probably encouraged animals such as pterosaurs and dinosaurs (and, later, bats) into the air. The Jurassic was when the first butterflies and beetles appeared, increasing the airborne menu – as well as the colourful artistry of nature. Life on our planet was always complex and breathtaking, but in the Jurassic it took on the colours of the startlingly rich world we see around us today.

Sudden Plethora

Although the most conspicuous inhabitants of the Jurassic, dinosaurs were far from the most significant. The first true mammals flourished at this time, as did the earliest modern amphibians, lizards and turtles. Many of these fed on new types of insect, including beetles, weevils, fleas, stick insects and moths.

When studying living organisms, researchers construct family trees, called phylogenies. They outline the relationships between living things, telling us which are closely related. In the Jurassic, the sudden appearance of new species burst across the phylogenies of life, like the spikes on a garden rake radiating outwards. Understanding what triggered this abrupt multiplication of species has long interested scientists. Despite fossil discoveries continuing apace since the early 1800s, there remain many unanswered questions about the origin of modern animal groups in the Jurassic. The middle of this time period is a mystery because very few fossils are preserved, due to the quirks of geology. Yet the geology itself hints at an explanation for the patterns of Jurassic evolution. The breakup of the supercontinent Pangaea probably isolated animal populations, allowing natural selection to change them in unique ways. The new habitats and climates created would have opened up novel niches for animals to crawl, fly and stampede into, as well as raise challenges to their survival. As is so often the case in the history of our planet, the very dynamism of the Earth herself was intrinsically linked to patterns of animal and plant evolution.

Archaeolepis – First Butterflies and Moths

Insects had been on the wing for tens of millions of years, but in the Jurassic the most captivating of them all appeared. The ancestors of butterflies and moths, such as the bee-sized *Archaeolepis*, first flitted through the Jurassic world. Butterflies are some of the most beautiful products of natural selection, forming not only a spectacular component of the ecosystem, but also inspiration for art and poetry around the world.

Archaeolepis was an insect that lived 190 million years ago, in what is now the south of England. It was just over a centimetre in length, and although known only from its delicate wings, they are enough to make this tiny creature important: it is the oldest known butterfly. These insects do much more than make our world more beautiful: they play an important part as pollinators and a food source for other animals. The study of fossils tell us that mimicry and camouflage – used extensively by butterflies and moths – are among the most ancient defences used by insects.

Butterflies and moths belong in a group called the lepidopterans. Despite being one of the most eye-catchingly diverse groups of insects in the modern world – found in every habitat on every continent except Antarctica – their fossil record is quite incomplete. Studies of their DNA suggest that the first lepidopterans evolved in the Late Triassic, but there are few fossils because their bodies are easily pulled apart after death and quickly decompose. *Archaeolepis* is therefore an incredible rarity, gently laid to rest in the Early Jurassic mud. Its wings were covered in microscopic scales, a characteristic shared by all lepidopterans. New butterfly and moth fossils have been found in China in the last few decades including whole bodies, as well as larvae, cocoons and leaves eaten by caterpillars.

Partnerships between butterflies and plants are common: the energy-rich sugars of nectar provide the fuel these insects need to flutter so extravagantly, and their larvae feed on leaves. There are over 180,000 species of moth and butterfly today, and most have close relationships with flowering plants. These intimate dependencies didn't become common until the Cretaceous, when flowering plants first appeared. In the Jurassic, lepidopterans used jaws to feed directly on plants, or deployed their sucking proboscis to gorge on sugary pollen drops exuded by gymnosperm plants.

The lackey moth, *Malacosoma neustria*, is just one of more than 160,000 moth species on our planet.

Having co-evolved for so long with flowers, many butterflies are crucial to pollination. From egg to caterpillar to adult, they are important prey for thousands of animals including mammals, birds and small reptiles. The magical process that turns a caterpillar into a butterfly has captured the human imagination for centuries, making them a symbol of transformation – most often positive, but sometimes terrifying.

Studying Beauty

Among insects, moths and butterflies have been lavished with the most attention from researchers. This is probably because they are not only inherently interesting, but also arrestingly beautiful. As a result, we know more about them than any other insect group. There are micro-moths barely bigger than rice grains, all the way up to the mighty Atlas moth (*Attacus*), whose wings are as wide as an open paperback book. Although renowned for their joyful fluttering, some species, such as skippers, can reach speeds of around 30 miles (50 kilometres) per hour. The hummingbird moth (*Macroglossum*) is not only fast, but also beats its wings in a blur like its namesake bird, allowing it to maintain position at a flower while feeding on nectar.

The monarch butterfly (*Danaus*) is famous for its incredible migrations. Thousands of them journey over 2,500 miles (4,000 kilometres) from Mexico to northern America each year with the changing seasons, guided by the position of the sun, the landscape or the electromagnetic field of the Earth. Most butterflies are daytime animals, but nocturnal species may even be guided by the stars. Moths generally fly at night, and scent is especially vital to them for communication and to find mates. Some lepidopterans even use sound to find mates and evade predators. Tiger moths confuse bats with their clicking chatter. Others can hear bats' ultrasound, and so escape by moving out of reach of hungry mouths.

Cunning Camouflage

Many organisms on Earth use camouflage and mimicry to protect themselves from predators. Butterflies and moths use it to blend in, or to stand out by resembling something much more dangerous. Their wings and bodies are covered with scales, like a shingle roof, and they give lepidopterans their name from the Ancient Greek, 'scaled wing'. These scales have a complicated surface that creates colours either through pigments or structures that diffract and change light.

The evolution of mimicry is a beautiful example of natural selection in action. When offspring naturally vary in pattern, those that are harder to detect or resemble an inedible species or predator have a survival advantage, and pass their genes and colouration to their offspring. Over time this can lead to wholescale changes in the appearance of a population of animals, creating the striking array of patterns we see around us today. Toxic butterflies and their larvae often advertise their danger using bold patterns and bright colours. Caterpillars are often green or brown to help them disappear among foliage, but adults may also resemble plants to avoid detection. This has been observed in fossil insects, including the ancient relatives of modern leaf-mimicking bush crickets and stick insects. Scorpionfly fossils found in Jurassic rocks from China mimicked the leaves of ginkgo, a common tree at that time.

One of the most iconic tricks butterflies use is having eyespots on their wings, mimicking the eyes of larger predators. They aren't alone in this ploy: the Chinese fossil lacewing *Oregramma* used almost identical eyespot patterns to fool Jurassic dinosaurs and pterosaurs 125 million years ago. Humans have even benefitted from the lessons of butterfly camouflage and mimicry, painting eyespots on their cattle and themselves to help protect from predation, for example in regions of Africa and India where lions, leopards and tigers often attack. By painting them on the rumps of the herd, or wearing a face mask on the back of their heads, people fool feline hunters trying to sneak up from behind. The study of camouflage and mimicry in animals has helped humans develop camouflage patterns for clothing, as well as new technology and materials based on the scales of lepidopteran wings.

Pterodactylus – First Vertebrate Flight

***Pterodactylus* was a pterosaur that lived in the Jurassic. These flying reptiles were the first backboned aeronauts to master life in the air. They were probably warm-blooded, and so versatile that they flourished throughout the Mesozoic, moving into every habitat and pioneering the life that birds and bats eventually followed.**

Pterodactylus lived 151 to 147 million years ago in Europe. It was quite small, with a wingspan of up to a metre (3 feet), similar to a peregrine falcon. Unlike birds, *Pterodactylus* was a pterosaur, and lacked feathers or a proper beak – although it may have had a small keratinous protrusion at the end of its snout. It had a long skull filled with teeth, and on top of its head was a double crest, sweeping up from the snout like a mohican. This may have been brightly coloured to attract mates and intimidate rivals. *Pterodactylus* fed on invertebrates and small animals such as mammals and lizards. Pterosaurs are the first vertebrates to evolve powered flight, and did so in a way that was unique compared to later dinosaurs and mammals.

Pterosaurs are often erroneously called 'pterodactyls', but *Pterodactylus* is just one of over 150 pterosaur species that lived in the Mesozoic. They were not dinosaurs but shared an ancestor with them. Although they flew, pterosaurs are not related to birds, but they were likely warm-blooded (endothermic) and had strong flight muscles in their arms. *Pterodactylus* was the first pterosaur to be identified in the late 18th century, but because there were no soft tissues preserved in the earliest specimens, it took some time for people to realize this was a flying animal. There are now multiple fossil pterosaurs with wing skin – known as membranes – preserved in the rock.

Pterosaurs appeared in the Triassic, and died out at the same time as the dinosaurs, 66 million years ago. Although often depicted as bat-like or lizard-like, they were neither. Whereas bat wings are thin and stretch between multiple fingers, like a webbed hand, pterosaur wings were tougher and supported on a single elongated fourth finger. The rest of the pterosaur hand was at the front edge of the wing, and may have been used when walking. There is evidence that pterosaurs may have been covered in small filaments called pycnofibres, which could have made them fuzzy to the touch. *Pterodactylus* and kin would have laid eggs, which were probably soft

This vintage depiction shows *Pterodactylus*; the first flying reptile, or pterosaur, ever found. It is one of over 150 pterosaur species now known from around the world.

and leathery. Baby pterosaurs (known as 'flaplings') have been found as fossils, but scientists are not sure how much care they received from their parents, or whether they were expected to fend for themselves soon after hatching.

We Have Lift Off!

The evolution of pterosaur flight – like that of insects, birds and bats – is still not well understood. It is difficult to trace how animals move from ground or tree-dwelling ancestors, to those borne aloft. Not only are the origins of pterosaur flight unknown, but the mechanics are also unclear. People once suggested that the largest pterosaurs in the Cretaceous could only stay in the air thanks to the different composition of the atmosphere at that time, but this is incorrect, because the difference was too small in relation to their body size to have had any effect. In the last thirty years, researchers have used the same mathematical equations engineers use to build efficient aircraft to study pterosaur flight. We now know these animals had strong wing and chest muscles, and once in the air they used the same properties of lift used by today's birds, flapping and soaring on air currents.

A more challenging puzzle is how pterosaurs launched themselves into the air from the ground. Most birds jump into flight, using their leg muscles to push upwards, whereas larger birds such as swans may run to pick up speed before taking off. Pterosaur legs weren't strong and they probably couldn't run, so for a long time it was thought they found high ground or climbed trees, leaping off and exploiting thermals to lift themselves into the sky. New research, however, tells us that pterosaurs were able to launch themselves using the very same muscles they use for flight, in their arms and chests. Like a rapid push-up, they thrust themselves upwards – known as 'quadrupedal launch' – before flapping away. This method is used by modern bats that land on the ground to feed, such as vampire bats.

Pterosaurs Through Time

Pterosaurs changed a great deal though time. The first groups to emerge in the Triassic had long tails and jaws filled with teeth. They were probably good climbers but less mobile on the ground, because their wing membranes were connected to their legs. Later in the Jurassic, some pterosaurs reduced their tails and wing membranes, and grew longer necks. Some developed enormous recurved teeth to catch fish, but by the Cretaceous, others were entirely toothless.

There is no limit to the foods that different pterosaurs ate: there were small, blunt-faced insect eaters and larger carnivores. Myriad fish-eaters emerged, from coastal foragers plucking prey from near the water's surface, to plunge-divers reaching deeper shoals. There were even shell-crushers and fruit- and seed-eaters. They filled every niche available to them on the planet.

Pterosaurs left their largest species to last. At the end of the Cretaceous, the azhdarchids were true giants, many standing as tall as giraffes and with wingspans over 10 metres (30 feet). They were probably predatory, and large enough to have eaten almost anything – including small dinosaurs. They walked, as many pterosaurs did, on all fours, with back feet flat on the ground and wings folded. With their limbs beneath them in this way, pterosaurs could traverse quite comfortably.

Pterosaur diversity declined towards the end of the Cretaceous. It's not clear why this happened, but what we do know is that when the asteroid slammed into the Earth 66 million years ago, no pterosaurs survived, and birds went on to take over the skies.

Beetles — Most Diverse Animals

In the Jurassic beetles suddenly infest the pages of deep time, munching their way across the landscape and leaving behind a glittering confetti of wing casings. They brought the Jurassic forests alive with their sight and sounds. Beetles are now the largest order of insects in the world, and fundamental to almost every ecosystem on the planet.

The origin of beetles is unclear due to a lack of fossils, but they almost certainly first appeared in the Early Permian. Scant fossils are known from the Triassic, mostly belonging to wood- and fungus-eaters. In the Jurassic they started to take on the many roles they play today, as detritivores, herbivores, carrion-eaters, dung-recyclers, predators, parasites and food for countless other animals. The first leaf, jewel, click and dung beetles appeared, as well as weevils. A little later in the Cretaceous, beetles were among the first insects to pollinate flowers, and many still fulfil this function today. Fossil beetles have even been found preserved in amber, coated in pollen like ancient confetti.

Beetles are found in every habitat except the poles. Some survive temperatures as low as –60 degrees Celsius (–76 degrees Fahrenheit) by entering diapause – a sort of hibernation – using energy reserves to stay alive, while a natural antifreeze prevents ice crystals forming in their bodies. At the opposite extreme, desert-dwelling species withstand a searing 50 degrees Celsius (122 degrees Fahrenheit), which would kill most other organisms. The beetle body plan has generally stayed the same since the Jurassic. They have hard outer casings called elytra, which cover their wings in a protective armour suit. These were originally forewings, but have been modified for this new purpose, and are the beetle's defining feature.

The range of beetle shape and size is staggering, from the Hercules, stag and rhinoceros beetles sporting their elaborate horns, to the elongated snouts of weevils. The largest beetles by weight are the larvae of the goliath beetle (*Goliathus*), at over 10 centimetres (4 inches) long and 115 grams (4 ounces). At the other end of the scale, the featherwing beetle is the width of two human hairs (325 microns, or 0.3 millimetres). They are often decked in dazzling costumes of colour, and use camouflage and mimicry with consummate skill. Beetles can communicate using pheromones, chemicals released to raise alarm, lay a trail towards food, or find a mate. They also produce

Beetles come in thousands of shapes and sizes, like these chafer (top left), darkling (top right) and stag (bottom) beetles. They are indispensable to ecosystems across the planet.

158

sounds, as with the squeak beetles, which signal annoyance by stridulation, rubbing parts of the body together to make an angry squeak.

God Loves Beetles

When asked what could be concluded about the divine from the study of natural history, the scientist JBS Haldane (1892–1964) joked that the creator had 'an inordinate fondness for beetles'. This was a fair observation: beetles comprise a quarter of all the animal species on Earth. There are over 380,000 species named to date, but researchers estimate 1.5 million or more exist, most of them as yet undiscovered.

There are lots of theories to explain the jaw-dropping number of beetles. They've been around for a long time, providing ample opportunity for natural selection. Their life cycle, in which adults and larvae are different and therefore don't compete for resources, may contribute to their success. Mammals have been a major predator of insects since their early evolution in the Jurassic, and many continue to feed on them. This could have driven beetles to adapt and produce new forms. For groups such as dung beetles, the appearance of the first megaherbivores, first in the Permian, then plant-eating dinosaurs in the Jurassic and Cretaceous, could have provided new food sources and habitats in the form of droppings, and changed how and where plants grow. The incredible flexibility of beetles to occupy lots of different niches makes them less prone to the extinctions that punctuate life on Earth.

Fossil Colours

Beetles are flamboyant: they can shine like metal, be iridescent, luminescent or even give ultraviolet signals. Some fossil insects are preserved in startling colours, but these may not have been their original garb in real life. Studies of the way colours change during fossilization show that the original colours shift into longer wavelengths. As a result, a violet-coloured beetle becomes bluer during fossilization, and a blue one a few shades greener.

It may seem a transient property, but colour can be built into living things in the way they are put together. These structural colours are produced by light interacting with the microscopic surface of a material. Most beetle wing cases are structurally coloured (as are many butterfly wings and bird feathers). Iridescence, when a material appears differently coloured depending on the angle from which it is viewed, is caused by multiple thin layers building up the surface, each reflecting light slightly differently. This can be detected in exceptionally preserved fossil insects, revealing that some ancient species were iridescent or brightly hued. Structural colours have appeared multiple times among beetle families. They may seem bright out of context but are camouflaged in their natural habitat, for example among the glittering leaves of a rainforest. Other colours warn predators or confuse them, providing time for escape. They signal to other beetles for mating, and even regulate temperature by absorbing or reflecting light.

Glittering Amulets

Beetles crawl throughout human culture, from Ancient Egyptian scarab amulets to beetle fighting, popular in Asia. Beetle wing cases have been used as jewellery or even incorporated into furniture. The ma'kech is a brooch from Mexico made from a living ironclad beetle (*Zopherus*), pasted with gemstones and attached to women's clothing with a small chain leash. Beetles can be major agricultural pests, but can also be our friends, such as the charming ladybirds that feed on aphids. Their varied roles recycling nutrients, burrowing in soil and providing food make them ecosystem lynchpins, vital to life on our planet today.

Archaeopteryx – Feathered Dinosaurs

Dinosaurs dominated the ecosystems of the Mesozoic. In the forests of the Jurassic flapped one species that, at a glance, looked for all the world like a weird raven. *Archaeopteryx* was covered in sooty black feathers but had teeth and clawed hands. This was the first of many fossils that have confirmed what scientists long suspected: dinosaurs gave rise to birds.

At the end of the Jurassic, there lived an animal that was part reptile, part bird, and 100 per cent astonishing. No bigger than a raven, it had wide feathered wings, and its long bony tail and slender running legs were also fringed with feathers, like outlandish cowboy chaps. Although somewhat familiar to us, peering back in time, this creature was something quite new to the world: a precursor of birds. *Archaeopteryx*'s long snout and mouthful of tiny teeth betray its reptile ancestry; at the end of each wing, a clawed hand that could still clutch and scratch. One of the most famous dinosaurs ever discovered, *Archaeopteryx* not only provided insights into an extinct world, but also into the genesis of our own. It told scientists that birds evolved from small, feather-clad Jurassic reptiles. They hadn't died out after all: our world is still filled with dinosaurs.

At least eleven *Archaeopteryx* skeletons have been found since 1861. They are sometimes preserved with a halo of feathers, like snow angels in the rock. Analysing the melanosomes in these feathers – the structures that hold coloured pigment – we know *Archaeopteryx*'s wings were predominantly black. It had a large brain and well-developed hearing and sight, and was active during the day, hunting beetles and small vertebrate animals such as lizards.

Although bird-like, *Archaeopteryx* and kin still had more in common with their dinosaur sisters than a modern magpie. Many groups of dinosaurs wore a feathery coating, and other small non-bird dinosaurs also had wings and a wishbone, or furcula. A recent fossil discovery from China, called *Anchiornis*, is now thought to be the oldest close bird relative. The ancestors of the modern birds (called Neornithes) probably appeared around ninety million years ago, in the Late Cretaceous. Animals like these are referred to as transitional fossils, or sometimes 'missing links' – an outdated term based on the idea of evolution as a straight line. The reality is much messier, with branches shooting in all directions and often ending

Since its discovery in 1861, the feathered dinosaur *Archaeopteryx* has provided incontrovertible evidence that birds evolved from dinosaurs.

162

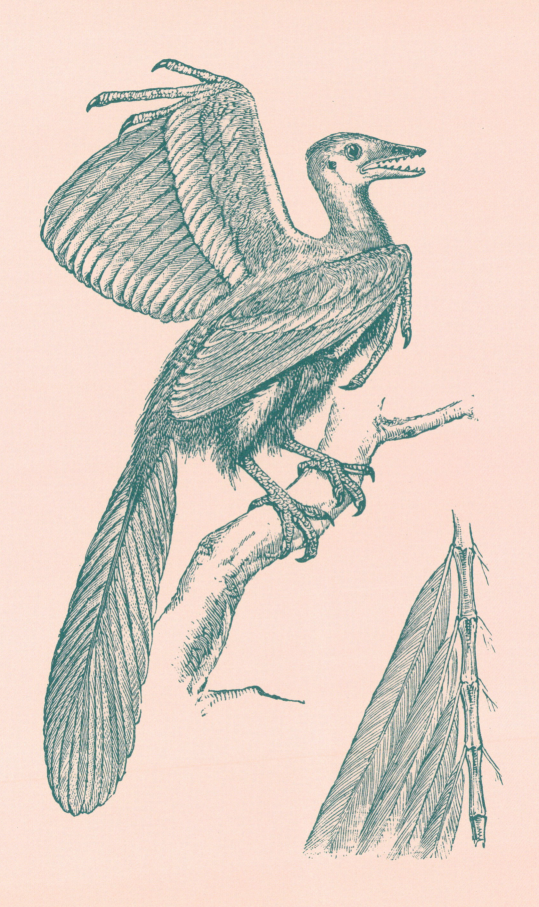

in extinction. It is true, however, that the anatomy of these fossils tells us more about the evolution of bird-like characteristics in deep time.

Whether *Archaeopteryx* could truly fly is still hotly debated. The structure of its feathers makes it capable of generating lift, a prerequisite for flight. The feathery legs of *Archaeopteryx* were probably splayed out and used as an aerofoil. Similar leg feathers are found in closely related non-bird dinosaurs such as *Microraptor*. If it did fly, *Archaeopteryx* was more likely to have been a burst flyer, like a pheasant, unable to stay in the air for long periods of time. Dinosaurs were the second vertebrate animal group to evolve powered flight, after the pterosaurs.

Dinosaur Dynasties

The fossils of Mesozoic reptiles have been uncovered by people around the world for many centuries. Among Europeans, the first dinosaur to be named by scientists in the 19th century was *Megalosaurus* from England, a large meat-eater that resembled *Tyrannosaurus*. Since that time, the focus of discovery has shifted from Europe and North America to China, South America and Africa. It is estimated that in recent years a new species of dinosaur is announced almost weekly.

There are three main types of dinosaur: the sauropods were enormous, long-necked, long-tailed giants made famous by the likes of *Diplodocus*; ornithischians include armoured animals such as *Stegosaurus* and *Triceratops*, as well as the duck-billed dinosaurs, and were predominantly herbivores; *Archaeopteryx* belongs to the third group, the theropods, as do other bipedal meat-eating dinosaurs. These three great lineages lived throughout the Jurassic and Cretaceous, being the most diverse and successful terrestrial animals during their 150 million-year existence. Yet only a few members of one branch survived the end-Cretaceous mass extinction: birds, which are theropods. We look to them for clues about the biology of their long-extinct relatives. To our knowledge, all dinosaurs were egg-layers, and they were probably warm-blooded. They had good vision, and feathered species may have used mating displays as birds do today. There are still many questions left to answer about these extinct creatures and the lives they led.

Feathered Forgeries

From time to time, people have claimed that the fossils of feathered dinosaurs such as *Archaeopteryx* are forgeries. Many of the arguments for this are based on a poor understanding of the process of fossilization, as well as underlying personal and religious beliefs that contradict the overwhelming evidence that evolution is the mechanism that has generated the diversity of life on Earth. However, forgery is occasionally a problem in palaeontology. Usually carried out for financial gain, there are instances where different fossil pieces have been stuck together, or features 'created' by forgers using human-made materials.

One of the most famous recent examples is *'Archaeoraptor'*, a so-called 'missing link' announced from China in 1999. Despite researchers casting doubt on its authenticity, it was unveiled at an international press conference as the latest amazing palaeontological discovery, and shared widely around the world. Closer examination of the fossil by experts revealed that it was a chimera, made of multiple different fossil dinosaurs and early bird relatives, arranged together to look like a single animal. With close scrutiny, fakes like these are usually detected. Technologies such as X-ray scanning are especially useful for scientists in revealing even the most skilful deception.

Cretaceous

The Cretaceous lasted eighty million years, making it the longest geological time period since complex life began. Global temperatures and sea levels both climbed, and the continents scattered, taking their cargos of life with them. On land, a most incredible new group blossomed: the flowering plants. They ignited a firework of diversity among all animals on land, a change so wholescale it was tantamount to ecological revolution.

The Cretaceous spanned from 145 million years ago to 66 million years ago, when it ended with a bang. Temperatures crept up throughout that time, reaching their maximum around 90 million years ago. The sea levels remained high, at around 110 metres (360 feet) above present-day levels, obscuring what would have been an otherwise familiar continental makeup. The South Atlantic and Indian oceans were born, and northern Africa, Arabia and Europe were submerged under the Tethys Sea, an ocean that eventually closed thanks to the northward drift of Africa. The proto-Caribbean Sea flooded parts of South America, while North America was divided in three by fecund seaways.

The Cretaceous is often abbreviated to 'K', from the German for chalk, *kreide*. This type of rock was laid down in deep layers across Europe at this time. The increased ocean circulation between the splitting continents enriched our seas with calcium, fuelling planktonic blooms. These plankton sank to the seafloor after death, forming enormous depths of this iceberg-white stone over millions of years.

This was an extraordinary time for life on our planet. Marine reptiles were common in the oceans, but birds were also exploiting the marine riches, including diving species such as *Hesperornis*, which resembled a cormorant. Antarctica was lushly forested, sporting populations of dinosaurs and mammals, their bones now hidden by ice at all but the fringes of the continent. Dinosaurs reached their heyday, and some became the largest land animals to have ever existed. Around them, life was being resculpted by the new opportunities that arose from the first flowers and fruit.

The end of the Cretaceous also marked the end of the Mesozoic and the reign of the non-bird dinosaurs. The mass extinction that snuffed out them and many other groups from existence was not as large as at the end of the Permian, but it was one of the five most devastating extinctions in evolutionary history. It put an end to many astonishing reptile lineages, leaving evolution to deal a new hand as the Cenozoic dawned.

Terrestrial Revolution

An extraordinary event happened between 125 and 80 million years ago. It is dubbed the Cretaceous Terrestrial Revolution (KTR), and marks a sudden burst of new types of birds, insects, mammals and reptiles. This world-changing moment can be tied directly to the evolution of the first flowering plants. Although dinosaur evolution is less influenced by this event, smaller reptiles such as squamates (lizards and snakes) flourished anew. The first members of modern mammal groups clambered through the boughs of fruiting trees, while around them pollinators busied themselves collecting a novel harvest of sweet nectar. For the first time in our planet's evolutionary history, the diversity of life on land surpassed that of the oceans.

Plants underpin food webs, and therefore have an incalculable effect on the evolution of other organisms. Flowering plants produced nectar and pollen rewards for insects, which in turn fed larger animals. The first bees buzzed through this sweet new world, and other social insects emerged, organizing themselves into colonies and building structures as protection against predators and parasites. These are now vital natural engineers of the living planet. The ancestors of modern mammals first appeared in the Cretaceous, but they remained overshadowed by their larger kin. It would take a mass extinction and the re-ordering of the living world to create the openings through which modern mammals finally slipped, growing into the dazzling forms we now recognize.

Asteroid Impact

In the uppermost layers of the Cretaceous lies one of the clearest of all geological boundaries: a layer of iridium, an element uncommon on our planet. It comes from the settled dust of a catastrophic asteroid impact 66 million years ago on the coast of the Yucatán Peninsula, Mexico. This is the K–Pg boundary, meaning Cretaceous–Palaeogene (formerly known as the K–T). The location of the impact, where the town of Chicxulub now sits, proved to be a lethal place to touch down. It is underlain by gypsum-rich rocks containing sulphate. As well as the immediate effects of the asteroid impact's shockwave, including forest fires and tidal waves, the sulphate was released into the atmosphere, forming sulphuric-acid rain. The ruinous effects dissolved ecosystems at every level, both on land and in the water.

The end of the Cretaceous is the most famous punctuation mark in our planet's evolutionary story. It is the mass extinction that erased almost all of the dinosaurs. The birds – their living descendants – survived, as did some mammals, and both have thrived in the last 66 million years. There were many other losses at the end of the Cretaceous: multiple species of backboned animal and photosynthesizing plants were walloped by the dust and cold of a nuclear winter. Pterosaurs, marine reptiles and the swirling ammonites all vanished forever. Those animals that remained founded a new iteration of food webs, in which mammals and birds took vital roles. It was the moment that shaped the world we know.

Archaefructus – Rise of Flowering Plants

In the Cretaceous the first flowers garlanded the Earth like a May Queen. One of these was *Archaefructus*, a humble waterside dweller with miniscule blossoms. From plants like this, nectar and fruit rewards were handed out to insects and other animals, forging new relationships in exchange for pollination and seed dispersal, and igniting top-to-bottom explosions in biodiversity. These amazing plants have underpinned civilizations around the world, and continue to provide food for us and many other animals.

The subtle, tiny flowers of *Archaefructus* are among the first on Earth. Flowering plants like this have transformed our planet – we couldn't live without them.

Around 125 million years ago, by the muggy bank of a forest pond, grew the soft green stems of *Archaefructus*. Tiny pointed flowers were arrayed around its apex. These swelled into miniscule fruits the shape and size of rice grains. The leaves feathered outwards at intervals down the stem, and the roots were simple, shaped for life in an aquatic habitat. This understated inhabitant of the Cretaceous world would have been easily trampled beneath the feet of a giant herbivorous dinosaur, stooping for a deep drink in the glittering pool. Humble it may be, but *Archaefructus* marks a multicolour evolutionary moment: it is one of the earliest flowering plants. Prior to this, our planet was devoid of flowers; afterwards, life was brought into floral technicolour.

Archaefructus comes from rock layers found in what is now northeast China. The exceptionally well-preserved fossils make it possible to identify its key place in the history of life. Flowering plants are called angiosperms, and today there are an estimated 400,000 species, making up 80 per cent of flora on Earth. The very earliest angiosperms can be hard to identify. *Archaefructus*, like others at this time, lacks obvious petals, sepals or other features we associate with flowers today. These early species lived mostly in or around lakes and streams. They differ from the other major group of seed-producing plants, the gymnosperms, by producing flowers and fruit. Many are able to develop from seed to flower in a single year – some, such as rockcress (*Arabidopsis*), can do it in a matter of weeks.

The rapid rise and spread of flowering plants fundamentally enhanced the productivity of ecosystems. Entire nutrient and water cycles were transformed. Throughout the Cretaceous, they replaced the old understorey of vegetation, and eventually ousted gymnosperms as the dominant trees across huge areas of our planet.

168

This triggered the Cretaceous Terrestrial Revolution, driving the evolution of almost every animal group on the planet today.

Feeling Fruity

Flowers and fruit have had a massive impact on evolution of other organisms, particularly through forming relationships with them for pollination and seed dispersal. The earliest land plants and algae reproduced using swimming sperm, whereas gymnosperms use the wind to carry their sperm in the form of pollen. Most flowering plants, on the other hand, rely on pollinators such as insects, birds and mammals to take pollen from flower to flower. By enticing them with colours, perfumes and rewards such as nectar and fruit, flowering plants use the pollinators in a task they couldn't achieve on their own. Although angiosperms appeared in the Early Cretaceous, it wasn't until after the end-Cretaceous mass extinction that they became particularly important to new species of modern birds and mammals. The size of fleshy fruits has fluctuated through time, coinciding with the growth and shrinking of forest habitats. At a site called Messel Pit in Germany, hundreds of well-preserved fossils have been found from animals living around 47 million years ago, in the Palaeogene. These include at least ten different species of fruit-eating mammals, showing how widespread this niche had become by that time.

Built on Flowers

Modern human agriculture is entirely dependent on flowering plants. From your first coffee in the morning to your evening meal, most of it is made from angiosperms. By far the most crucial to us are the grasses, which include barley, maize, rice, oats and wheat. Then there are the gourds (which includes pumpkins and squashes), the rose family (which includes most fruit trees, such as apples and plums), the nightshade family (that holds the peppers, potatoes and tomatoes) and the citrus fruits. Many of them have multiple uses, such as the coconut (*Cocos nucifera*), which can be eaten but also generates fibres for clothing, building materials, utensils and jewellery. As well as direct foods for us, flowering plants are also the primary feed for our livestock.

Evolutionary psychologists have suggested that some flowers may have been given an evolutionary advantage, thanks to humankind's eye for beauty. During our earliest agricultural practices, flowers we found attractive might have escaped land clearance, being left to grow simply because we enjoyed having them around. Humans have been growing ornamental flowers for at least five thousand years – pollen has even been found in the graves of our close kin, the Neanderthals, hinting that they also appreciated blossoms. Such aesthetic qualities may have contributed to the spread of some flowers at the expense of others, influencing their evolution. Even the paper this book is printed on is probably made from angiosperms.

Melittosphex — Origin of Bees

Melittosphex was one of the first bees on Earth. These crucial pollinators evolved from wasps, their anatomy emerging in step with flowering plants. Starting out small, bees have since busied themselves into the complex, social insects we know today. Fuelled by nectar and producing rich velvet honey, they are vital for human agriculture, but face an uncertain future – with potentially devastating consequences for us all.

Floating in a golden sea of amber, *Melittosphex* is the fossil of an ancient bee. Only about 3 millimetres long, this tiny insect lived in a hot tropical forest 100 million years ago. It had a heart-shaped head and long, delicate legs. Small branched 'hairs' on its hind legs are thought to have been used to collect pollen – indeed, pollen grains are still visible dotting its legs and head. *Melittosphex* is one of the world's first pollinators.

Bees are herbivores, but they evolved from among the carnivorous wasps. *Melittosphex* has a mixture of features from both groups, making it a crucial transitional fossil linking them together. Bees and wasps are both hymenopterans, a group of insects that also includes ants and sawflies. Hymenopterans may have emerged as early as the Permian, but their fossil record is scanty. The fossil of *Melittosphex* tells scientists that by the Cretaceous, the features we associate with bees had already appeared, shedding light on their evolutionary path. Many of the first bees were extremely tiny, which is to be expected as the first flowers were also small. Fossil flowers in the same amber deposits as *Melittosphex* are only 1 to 6 millimetres across. The origin of pollinators such as these is intrinsically tied to the evolution of the first flowering plants.

Although most of us are familiar with honey-producing bees and bumblebees, there are over 16,000 known bee species found on every continent except Antarctica. Some are as small as *Melittosphex*, such as the genus *Perdita*, whereas Wallace's giant bee, or the rotu ofu (*Megachile pluto*), can reach almost 4 centimetres (1½ inches) in length. It is often said that bees should not be able to fly because they don't conform to the laws of aerodynamics, but this is a misconception arising from a misunderstanding of the mechanism they use to become airborne. We now know they use extremely fast, short wing strokes, beating at around 230 flaps per second. Some bees have stingers that they use in defence. Unlike wasps, bees are generally

Bees are one of the most important pollinators, a relationship with plants that reaches back to the Cretaceous.

172

associated with positive traits such as industriousness and co-operation. In Egyptian mythology, bees were born from the tear-drops of the sun god Ra.

Precious Pollinators

Bees are the world's most important pollinators. The first flowers were likely fertilized by other insects, such as beetles. As bees evolved their unique relationship with flowers, they have been moulded as designated couriers for plant reproduction. To fulfil their task they have long tongues to extract nectar, and special pollen-carrying leg 'hairs'. There are even nocturnal bees, which feed on blossoms that only produce nectar at night. They detect suitable flowers by their scent or ultraviolet patterns on the petals. Honey bees return to the hive to communicate the location of food sources using a 'waggle dance', telling the rest of the hive in which direction to take flight.

Honey is made by bees from nectar. Bees keep the nectar in a special honey stomach, where it is partially broken down by enzymes. In the hive they regurgitate it, and hive bees digest it further and remove some of the water content. The final product is stored in a wax honeycomb. The hexagonal lattice of honeycomb is created by the tension between the tubes, which are originally circular. These honey reserves are resistant to bacteria and mould, and keep the hive alive through winter.

Humans have been bee-keeping and collecting honey from wild bee nests for centuries. The oldest depiction of honey gathering comes from eight thousand-year-old rock art in the Cuevas de la Araña caves in Spain. China is now the world's largest honey producer, harvesting 1.9 million tonnes annually, a quarter of all commercial honey production.

At least a third of human agriculture is based on cultivating flowering plants that require pollination. Most of this is carried out by bees, both wild and domesticated. These hard-working gatherers are fast disappearing from our planet, thanks to a lethal combination of pesticides, disease and a decrease in wild flowers. Climate change is exacerbating these factors. It has been forecasted that if we don't hasten our efforts to save bees from extinction, their disappearance will be a catastrophic loss for both humans and the ecosystem at large.

Amber's Dark Side

Melittosphex was found preserved in fossilized tree resin, called amber. There are several places globally famed for their amber deposits, each dating to a different period in Earth's history. Amber from the Dominican Republic comes from forests around 23 million years ago, whereas Baltic Amber is around 44 million years old. *Melittosphex* was found in amber from Myanmar (Burma), which captures an exceptional slice of the ancient Cretaceous world in its orange time capsules.

However, Myanmar amber often comes from mines that are involved in the ongoing civil war in the region. Many mines have been seized in violent takeovers involving human rights abuses, and the mine workers may endure dangerous working conditions. This makes studying Myanmar amber a difficult ethical issue for palaeontologists. In recent years, highly publicized Myanmar amber fossils have caused the price of this resource to skyrocket. Illegal sales often fund the continuing violence. Although it may provide unparalleled insights into the past, the human cost is immense, and an increasing number of researchers are calling for a halt in research on Myanmar amber fossils until the situation in which it is obtained improves.

Repenomamus — Dinosaur-eating Mammals

Repenomamus was one of the largest mammals known from the time of dinosaurs. This stocky badger-sized creature lived in China in the Early Cretaceous. One astonishing specimen was found with stomach contents intact: the remains of a baby dinosaur. This is one of many fossils upturning our ideas about mammal evolution in the Mesozoic.

Around 130 million years ago, an animal called *Repenomamus* padded through the undergrowth. It looked like a badger, covered in fur with a stocky body and sharp canines. *Repenomamus* grew up to 14 kilograms (31 pounds), making it one of the largest mammals in the Mesozoic. It belonged to an extinct group called the gobiconodontids, the first specialist meat-eating mammals. Although they mainly fed on smaller vertebrates such as lizards and small mammals, incredible evidence from China reveals these hungry opportunists were not above feasting on dinosaur hatchlings, turning our preconceptions about ancient food webs on their head.

Although people associate the Jurassic and Cretaceous with reptiles, mammals were thriving alongside them. They are part of the enormous group called the synapsids, which split from their shared ancestor with reptiles over 300 million years ago. By the Late Triassic, most of the synapsid lineages had died out, leaving only mammals. Until recently, it was thought that mammals remained mouse-sized in the Jurassic and Cretaceous because they were dominated by the giant reptiles they lived alongside. We now know, thanks to fossils such as *Repenomamus*, that this is not true. Inside its stomach were the remains of a juvenile *Psittacosaurus*, a common plant-eating dinosaur. Although it's not possible to say if it actively hunted or simply scavenged its impressive meal, this fossil confirms that mammals were much more ecologically diverse at this time than anyone previously suspected.

There are three main mammal groups alive today: the placentals, the marsupials and the monotremes (platypus and echidna). They share a common ancestor as far back as the Triassic. All mammals are warm-blooded, produce milk and have a body covering of hair. They have complex-shaped teeth that, unlike in other backboned animals, are usually replaced just once by an adult set that must last them a lifetime. Our world is now replete with mammals of

Repenomamus was a carnivorous mammal the size of a badger that ate baby dinosaurs.

all shapes and sizes, but all of these stem from just a few survivors of the end-Cretaceous mass extinction. Prior to that, myriad families shared the dinosaurian world, including *Repenomamus*. Most of these disappeared along with their reptile cohabitants – leaving the Earth free for the ancestors of modern mammals to take over.

Legacy in the Senses

In the Triassic the very first mammals were extremely small, and probably nocturnal. Becoming small nighttime specialists has left a permanent legacy in mammal biology. Smaller animals lose more heat than larger ones because of the higher surface area to volume ratio, with body heat being lost through the surface of the skin. The first mammals compensated by developing an insulating cover of fur, and their metabolisms sped up – part of the reason mammals are warm-blooded today. We know the first mammals were nocturnal due to the eye structure and genes of living mammals. Mammals today have very few of the light-sensitive structures in their eyes called cones, which are used for daytime vision and colour perception. These were not needed by their nocturnal ancestors, and so the cones and genes associated with them were lost. As a result, most mammals are now colour-blind, with only a few lineages (including ours) evolving alternative ways to detect colours.

Being nocturnal probably led to the development of enhanced senses, including sensitive hearing and scent detection. Mammals can hear an enormous range of sound from the ultra-high frequencies detected by bats, to the lowest vibrations of elephant conversations. Mammals also rely on scent for communication, and their whiskers and fur are attuned to touch, ideal for navigating in low light. All of these changes probably contributed to the increase in brain size seen in mammals starting in the Jurassic, as they adapted to interpret the increasing sensory information coming from their surroundings. Without this, it is unlikely we would have the same range of species we have today, from the tiniest feisty shrews, to deep-ocean behemoths such as the blue whale.

Unexpected Diversity

In the last twenty years new fossils have overturned our picture of mammals in the time of dinosaurs. These creatures ranged from paperclip-sized nibblers to bulldog-esque meat-eaters. Specialist climbers navigated the treetops with long grasping fingers, accomplished swimmers dived for aquatic insects and fish, and mole-like diggers feasted on worms. There were even gliders, like today's flying squirrels, using outstretched skin flaps to aerofoil between the trees. Most of these fossils come from China, where they are preserved in astounding detail. They reveal that mammals in the Mesozoic were almost as ecologically diverse as similar-sized animals today.

While all of this amazing diversity occurred in multiple mammal groups in the Jurassic and Cretaceous, the ancestors of modern mammals remained unremarkable. As the Earth's continents broke up, they were separated, and after the mass extinction, each population founded unique lineages in different parts of the world. The Afrotherian mammals, for example, including elephants, golden moles, manatees and hyraxes, trace their ancestry to the Afro–Arabian continent. The laurasiatherians, comprising hedgehogs, whales, hoofed mammals, carnivorans and bats, have a common ancestor in the northern hemisphere. Marsupial mammals are found in Australia and South America. This is an example of how intimately the planet's geology is tied to its biology, creating distinct patterns in life on Earth.

Argentinosaurus – Largest Land Animals

It is a cliché to say that dinosaurs shook the Earth, but some must have made this phrase a reality. By the Cretaceous, many types of dinosaur were enormous, but the largest were animals such as *Argentinosaurus*: the sauropods. With their long necks and tails, and colossal bodies, everything about them screams giant. Theirs was a dinosaurian utopia, packed with creatures so extraordinary that they have captured human imagination for centuries.

The sauropod *Argentinosaurus* lived between 96 and 92 million years ago, in the Late Cretaceous of what is now Argentina. At over 30 metres (nearly 100 feet) from snout to tail tip, scientists estimate this creature weighed as much as nine elephants. *Argentinosaurus* had a long, snake-like neck and preposterously tiny head. Its bulky

body was held aloft on sturdy legs, with a long tail swept behind it. Sauropods are famed for their museum-dwarfing scale. Perhaps the best-known species are *Diplodocus* and *Brachiosaurus*, but these were quite modest in size by comparison. *Argentinosaurus* was a titanosaur, a type of sauropod that includes the biggest animals to ever walk the face of the Earth.

Most of the biggest dinosaurs lived in South America. In the Cretaceous, the continent was separated from Africa to the East, and North America just across the equator. *Argentinosaurus* wandered among a network of rivers that braided through the plains. A verdant blanket of conifers coated the hillsides, providing fodder for these ceaseless eaters. Their feet were unlike an elephant's, instead possessing claws that curved to the side. Although we know sauropods reached colossal size, their mass can be only be estimated because

Plant-eating, long-necked sauropod dinosaurs like *Argentinosaurus* were the largest land animals that have ever lived on Earth.

nothing like them remains alive today. Other dinosaur giants include *Patagotitan*, *Xinjiangtitan* and *Dreadnoughtus*. All of them were plant-eaters, a true testament to the fecundity of the Cretaceous ecosystems that sustained them.

Not Too Big to Walk

There has been much debate about how such gargantuan animals as sauropod dinosaurs could walk on land. Most of the largest animals we know today are aquatic, and the water supports their body weight. It was once suggested that sauropods must be aquatic, using their long necks to keep their heads above water to breathe. We now know that although gravity presents biological challenges for bigger animals, nature has repeatedly found ways to surmount them. Dinosaurs such as *Argentinosaurus* were indisputably land dwellers. One of the ways they overcame the problem of extreme weight was by evolving air sacs and hollows in their bones, making them much lighter. For this reason, we can't use body mass estimates from mammals to understand how much sauropods weighed, because their bones are not structured in the same way.

Like other dinosaurs, sauropods were probably warm-blooded, and needed to eat a lot to survive. They stripped reams of leaves from branches quickly, swallowing them with little to no chewing. Sauropods laid huge clutches of relatively small eggs, probably no bigger than a football. Their young were little, but we know from studying the microscopic structure of their bones that they grew exceptionally quickly for the first decade or two, gaining as much as 2 tonnes per year, and continuing to grow slowly for the rest of their lives. As adults they were safe from meat-eating theropod dinosaurs, which is probably what drove the evolution of their enormous body sizes. Other dinosaurs adopted different strategies, including body armour such as spikes and thick skin plating, and living in herds. Similar defences can be seen today in large mammals, such as rhinos, musk oxen and wildebeest.

Dinosaurs are famed for being enormous, and although this is cited as a measure of their success, being big was simply another strategy for survival. They weren't very accomplished at being small – a niche well occupied by other lineages of animals including lizards, amphibians and mammals. Only the ancestors of birds also reduced body size, one of the adaptations that probably helped them survive the extinction at the end of the Cretaceous.

**The Mighty
Have Fallen**

The Late Cretaceous world hosted the most iconic dinosaurs: creatures such as *Tyrannosaurus rex* and *Triceratops* in North America; small hunters such as *Velociraptor* in Mongolia and China; and the enormous titanosaurs of the southern hemisphere. Dinosaurs thrived for over 150 million years, but almost all of these animals disappeared at the end of the Cretaceous, when an asteroid hit the Earth in what is now the Gulf of Mexico.

Thanks to our fascination with them, researchers have spent a great deal of time trying to understand the extinction of the dinosaurs. Some studies have suggested they were already in decline by the end of the Cretaceous, based on apparent drops in their numbers seen in the fossil record. However, the patchy distribution of fossils around the world means this interpretation is far from clear-cut. One thing is certain: they did not live beyond the end of the Cretaceous.

Birds did make it through, though, and continue to prosper. This was probably due to a combination of factors including their small size, which meant they required less food and could shelter more easily from adverse conditions following the impact. Possessing feathers, they may have been able to withstand the cold of the 'nuclear winter'. Many of the successful groups were diving birds, perhaps because of the flexibility provided by their diet. Birds often care for offspring after hatching – unlike sauropods, and perhaps other dinosaur groups – and this could also have enhanced their survival through one of the most difficult periods in the history of life.

CENOZOIC

Cenozoic means 'new life', but this is somewhat of a misnomer. Non-bird dinosaurs and their reptile cousins disappeared forever, but almost all other animal groups have their beginnings before the Cenozoic dawned. However, it was a fresh phase, when many of the characters we know today began their journey into the limelight. The Cenozoic is populated by creatures we know, and those that seem to have stepped from the pages of science fiction. The time span from the end-Cretaceous mass extinction to 2.6 million years ago used to be called the Tertiary, but this term has been replaced by the three periods of the Cenozoic: the Palaeogene, Neogene and Quaternary. We exist in the last slice of the final one of these delineations, the most recent word in a book that is still being written.

Our continents gradually took their current positions, and although the Earth dried and cooled throughout the Cenozoic, fluctuations turned up the heat for brief moments, driving changes in habitats and animal life. India sailed hastily north across the Indian Ocean and careened into Asia fifty million years ago. Where they slammed together, the Himalayas were lifted sky-high. It is thought that erosion of those peaks changed the global carbon cycle, triggering climate cooling. North and South America remained separated by the Strait of Panama until only a few million years ago, keeping their living populations isolated. When the Isthmus finally connected them like a geological handshake, their unique wildlife intermingled: an event called the Great American Biotic Interchange (GABI), creating an unusual mosaic of marsupial and placental life across these continents. The geological changes also impacted ocean circulation, resulting in strong Pacific and Atlantic currents that pulled warmth away from the once-lush Antarctic. Soon the southern continent was clothed in ice.

The devastation caused by the end-Cretaceous asteroid impact decimated the mammals and birds that had previously flourished, but life replenished at breakneck speed. Mammals were, at first, a jumbled mishmash of creatures of confusing heritage, but soon they took distinct paths, including carnivorans, whales, hoofed mammals and monkeys, along with other groups now extinct, such as tiger-like creodontans and elephantine gomphotheres. The first and only flying mammals, bats, flap suddenly into the fossil record, their origins obscure. Birds, meanwhile, fared well: the earliest penguins evolved in the southern ocean margins, while South American 'terror birds' stood taller than two ostriches. Although plants were damaged in the short term by the extinction event, flowering plants were surprisingly resilient. Animals that relied on them for food may have had an advantage. Grasses are the pivotal group of the Cenozoic, shaping ecosystems and putting selective pressures on those that grazed upon them, as well as underpinning human agricultural civilizations.

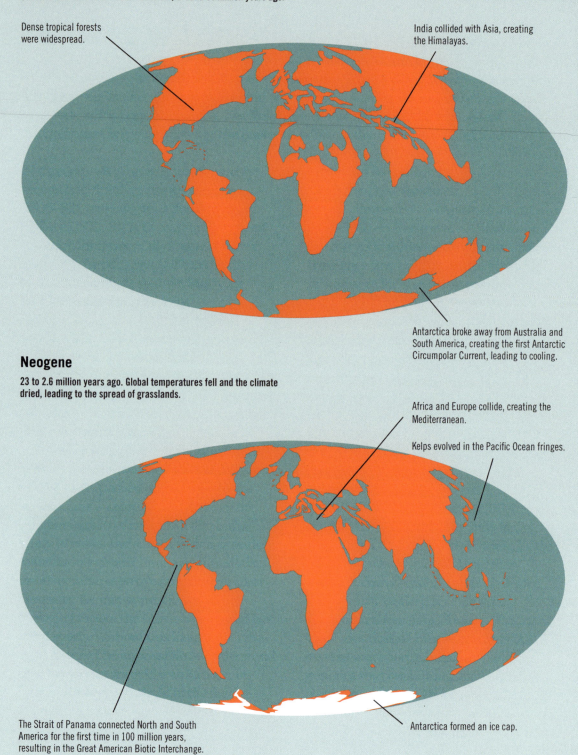

Palaeogene

66 to 23 million years ago. The planet heated up, reaching the PETM (Palaeocene-Eocene Thermal Maximum) around 55 million years ago.

Dense tropical forests were widespread.

India collided with Asia, creating the Himalayas.

Antarctica broke away from Australia and South America, creating the first Antarctic Circumpolar Current, leading to cooling.

Neogene

23 to 2.6 million years ago. Global temperatures fell and the climate dried, leading to the spread of grasslands.

Africa and Europe collide, creating the Mediterranean.

Kelps evolved in the Pacific Ocean fringes.

The Strait of Panama connected North and South America for the first time in 100 million years, resulting in the Great American Biotic Interchange.

Antarctica formed an ice cap.

Ice ages in the Quaternary

2.6 million to 11,700 years ago. Lower global sea levels exposed more shorelines and islands.

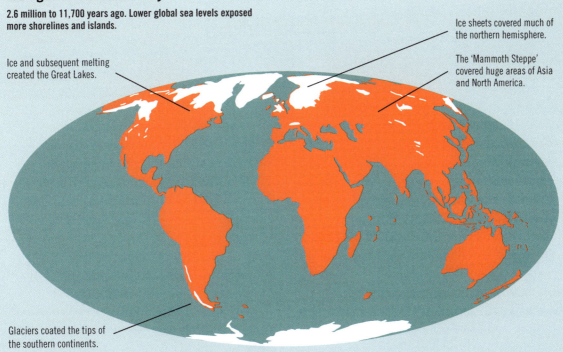

Ice and subsequent melting created the Great Lakes.

Ice sheets covered much of the northern hemisphere.

The 'Mammoth Steppe' covered huge areas of Asia and North America.

Glaciers coated the tips of the southern continents.

Anthropocene

Present day. Earth faces unprecedented rapid climate change caused by humans. Global habitat destruction and loss of biodiversity. Sea level rise threatens small islands and coastal communities.

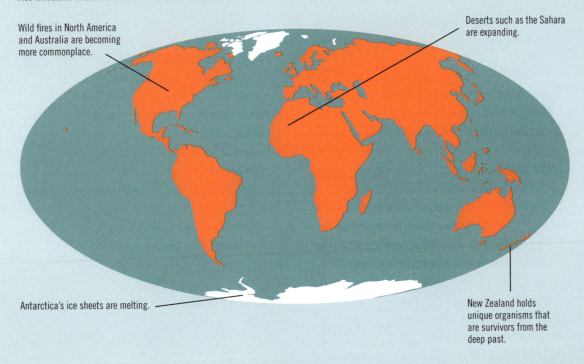

Wild fires in North America and Australia are becoming more commonplace.

Deserts such as the Sahara are expanding.

Antarctica's ice sheets are melting.

New Zealand holds unique organisms that are survivors from the deep past.

Palaeogene

The Palaeogene lasted 43 million years, during which life recovered from mass extinction, and the animals we recognize today began their journey. The evolution of mammals and birds dominates our focus, but the smallest organisms on land and sea were also undergoing changes – being shaped by, and shaping, our recovering planet. This hothouse world teaches us about extreme climate change, and how tightly the living world is interwoven.

The Palaeogene lasted from immediately after the end-Cretaceous asteroid impact 66 million years ago to 23 million years ago. The Earth warmed at first, sending out fingers of jungle all the way to the polar regions. Around 55 million years ago, this heating peaked in a major hothouse event that shaped life on land and sea. Dense tropical forests dominated many landscapes for a time, but by the end of the Palaeogene they were in retreat, as a global cooling trend took hold.

The Atlantic Oceans continued to broaden their spans, and sea levels fell, exposing coastlines easily recognizable as our own. India steamed across the Indian Ocean from its former home among the southern continents, ploughing into Asia to form the Himalayas. This collision continues to the present day, raising these mountain peaks and the Tibetan Plateau by half a centimetre every year. As the summits reached for the sky for the first time in the Palaeogene, they were battered by rain and wind, drawing down carbon from the atmosphere. As abruptly as the temperatures had risen, this caused them to fall again, and ice was soon locked into glaciers at the poles and mountain summits.

Gathering itself together at the start of this time period, nature set about rebuilding the animal world from the surviving lineages. Some archaic mammals, the first members of the groups we know today, moved out of the niches they inhabited so expertly in the time of dinosaurs and explored new ways of life. Birds soared through a sky free of pterosaurs, and swam in a sea where marine reptiles had disappeared – although sharks soon become the main predators of the ocean. Penguins proliferated in the southern oceans. In response to the initial heat of the Palaeogene, many land animals reduced in size – the earliest horses, for example, were no bigger than a dog. This dwarfing trend is due to 'Bergmann's rule', which states that body mass increases in cool climates and decreases in warmer climates. There are many possible reasons for this: the increase in carbon dioxide which had raised temperatures may have lowered the nutritional value of plant foods; forest habitats and scarce resources are also best suited to smaller bodies. As the climate cooled off again, giants emerged once more.

Circumpolar Current

Everything that happens on our planet is linked intricately together. The shifting of continents not only altered animal and plant life, but also rerouted ocean currents, affecting climate patterns around the world. In the Palaeogene, Antarctica finally drifted away from its closest neighbours, Australia and South America. The exact timing of this event is uncertain, but between twenty to forty million years ago, the Tasmanian Seaway and the Drake Passage opened up, providing an uninterrupted flow of water around the southern hemisphere for the first time in perhaps as much as a billion years. This is called the Antarctic Circumpolar Current, and it shaped our climates and subsequent life on Earth.

The Antarctic Circumpolar Current flows clockwise around the continent of Antarctica and is the largest ocean current in the world. Its presence erects a marine barrier, preventing warm water nearer the equator from drifting down and thawing the southern continent. As a result, Antarctica remains locked in the freezer, the coldest continent on the planet. The formation of the Antarctic Circumpolar Current in the Palaeogene was one of the reasons that Earth's global hothouse climate cooled back down again, a trend that continued from that point onwards.

Although the freezing of Antarctica put an end to the rich forest ecosystems that once thrived there, it also gave life to new ocean riches. Where warm seawater abuts the cold currents north of the Antarctic continent, a soup of nutrients and plankton is stirred up from the ocean floor. This creates deliciously rich seas supporting a burgeoning food web. Today, swarms of krill gather to gorge themselves, and feeding on them are fish, seals, penguins, seabirds and whales. Those that retain a tie to land still haul themselves onto Antarctic shorelines to breed and nest in this most extreme of environments. The massive southern ice sheets not only had a profound effect on marine life, but also on the history of human endeavour. Multiple expeditions have set out to explore their secrets, and although the ice sheets remain virtually uninhabited, the people that do dwell there, in small settlements clinging to the fringes, do so almost entirely for the purposes of scientific research.

Barylambda – Mammal Grab Bag

As nuclear winter gave way to the Cenozoic spring, mammals emerged from their burrows to reclaim the world. A mere six million years into the Palaeogene, *Barylambda* hefted through the tropical undergrowth. It was the first of the mammal megabeasts, blazing a trail for future giants. Although their family relationships remain enigmatic, animals such as *Barylambda* are among the precursors of the mammals we know today.

Barylambda lived in Colorado and Wyoming in the United States 60 to 57 million years ago. The animal itself was bear-like: thick-set and flat-footed. It had a small head and exceptionally chunky tail, swaying slowly as it padded through the dense forests of the Palaeogene. We know from the shape of its teeth that it ate plants, perhaps rearing onto its hind legs to strip higher branches. *Barylambda,* unlike most mammals from the Palaeogene, is known from relatively complete fossil skeletons, giving us an unprecedented picture of this pioneer herbivore. At around the size of a pony, it was one of the largest mammals at the time, and certainly the largest since the Triassic.

Only a few mammal groups made it through the devastation of the end-Cretaceous mass extinction. This was probably thanks to a combination of their biology (warm-blooded, milk-producing and coated in fur); their behaviour (caring for their young, and burrowing); and also a dose of luck. With the large reptiles removed from the ecological picture, their seats were up for grabs by whoever could adapt to fill them. Once diminutive, some mammals quickly grew big-boned and sturdy, such as *Barylambda*. Beneath its furred skin, *Barylambda* had an elephantine skeleton, but it belonged to no living mammal group. It was a pantodont, a group that thrived in the Palaeogene but became extinct around 38 million years ago. They were among the first large herbivores in the post-dinosaur world, leaving their fossils across North and South America and Asia.

Within the first ten to twenty million years of the Palaeogene, all of the major mammal groups we know today were established. By the end of the Palaeogene, the largest land mammal of all time, a type of rhinoceros called *Paraceratherium*, lived in Eurasia. It looked like a cross between an elephant and giraffe, standing at 5 metres (16 feet) at the shoulder. It had a long neck holding up a skull nearly as large

The first mammals to evolve in the Palaeogene, like *Barylambda*, took advantage of the empty ecological niches that surrounded them, founding the current 'age of mammals'.

as a person. However, the path of mammals from opportunistic impact survivors to modern species and mighty giants is far from clear.

Making Sense of Old Bones

Sometimes the palaeontologists studying fossil animals just can't figure out what they are. This happens when too little of an animal remains to reveal its place in the family tree, or because the animals themselves lack distinguishing features that unequivocally separate them from one another. In the Palaeogene, one such group is the 'condylarths': a jumble of animals believed to include the ancestors of hoofed mammals such as horses, hippopotamus, deer and cows. Groups such as these are known as 'wastebasket' groups, referring to the way in which a mishmash of fossils are thrown together, like litter chucked into a bin. Some members of wastebasket groups might be closely related, but others are probably not. Researchers continue to struggle to make sense of their relationships, occasionally making a new discovery that pulls one of these confusing creatures from the mire.

Another wastebasket grab bag in the Palaeogene are the creodonts. These beasts were the most fearsome hunters on Earth for over fifty million years. They looked superficially similar to the carnivores we know today – such as dogs, cats and bears, which belong to the order Carnivora – but are only distantly related. Creodonts included the largest carnivorous mammals ever known, such as *Sarkastodon* from China and Mongolia, which reached 3 metres (10 feet) in length. Like pantodonts, and many other early branches of mammals that emerged after the end-Cretaceous extinction, it isn't clear why creodonts became extinct. Eventually, however, they were replaced by modern carnivorans, which are now the most successful carnivores on almost every continent.

Whales and Bats

They may seem as different as two groups can get, but whales and bats have more in common than you might think. Each has evolved to live in ways no other mammal has before. Both have developed super-high frequency hearing to hunt and navigate their world, and incredibly, they've done so via the same gene mutations, making it possible to echo-locate. Their evolutionary journeys reveal the astonishing mechanisms that underlie natural selection.

Thanks to a wealth of jaw-dropping fossil discoveries, we know that whales and dolphins evolved from among the hoofed mammals.

Approximately fifty million years ago, their land-living ancestors began spending increasing time in water, and natural selection eventually adapted their bodies to the aquatic environment: streamlining the body; reducing and losing the hind limbs; adapting the forelimbs into flippers; flattening the tail (called a fluke). These changes can be tracked in the fossil record, leaving no doubt as to how natural selection changed the mammal body for a life beneath the waves.

In contrast, due to their delicate bones we know almost nothing about the origins of bats. The oldest fossil is *Icaronycteris*, which lived 52 million years ago, and it is already fully 'batty' in anatomy. Bats evolved flight in a novel way, developing a membrane between outstretched fingers. They are now the second-largest order of mammals, and play a crucial role as pollinators and seed dispersers, as well as benefiting humans by eating insect pests and providing fertilizer in the form of droppings. Humans rarely came into contact with bats in the past, but as populations have expanded and we increasingly rub shoulders with these incredible little animals, we have been exposed to their pathogens. Bats are believed to be the source of Covid-19, cause of the global coronavirus pandemic that began in late 2019.

Waimanu – A Bird's World

In the Palaeogene, a shallow sea washed over New Zealand. Swimming in it was an extraordinary animal called *Waimanu*, one of the world's first penguins. Their rare bones help piece together the story of the birds which survived mass extinction. They now form stunning flocks of species, from giant South American killers, to the most prolific bird on Earth: the clucking chicken.

Waimanu was the size of an emperor penguin, with short sturdy legs and powerful little wings. It sported a long narrow beak, and stood more or less upright, with webbed feet perfect for paddling. *Waimanu*, and the slightly smaller *Muriwaimanu*, are the oldest penguins known in the fossil record, and dived in the ocean over what is now New Zealand 60 million years ago. New Zealand was already isolated from Australia and Antarctica at this time, and this isolation led to the unique animals and plants that give this country a character unknown in any other part of the world today. *Waimanu* is important because of what it tells us about the evolution of birds. By examining fossils like this, combined with DNA analyses of living bird species, researchers know that modern birds emerged around the time of the end-Cretaceous extinction. The name *Waimanu* comes from the Māori words for 'water bird', and it seems that adaptations for a life at the water's edge helped some birds endure the worst that life could throw at them.

There are around twenty penguin species today, almost exclusively living in the southern hemisphere. Despite being associated with Antarctica, most penguins inhabit more northerly coasts, including one species on the Galapagos, near the equator. Although there are hundreds of seabird species, penguins are among the few that have committed fully to a life underwater and on land, becoming entirely flightless. Graceful and lightning-fast in the water, they also spend around half of their lives resting and breeding on beaches, rocky shores and icebergs. The largest penguin is the emperor penguin (*Aptenodytes forsteri*), which stands at just over 1 metre (3 feet), while the smallest is the little blue penguin (*Eudyptula minor*), at only 33 centimetres (13 inches). There are many extinct species, including some true giants. Just 23 million years after *Waimanu*, the giant penguin *Pachydyptes* also fed in New Zealand's waters. At around 1.6 metres (5 feet) tall, it is one of the largest and heaviest

Todays' penguins, like this vintage engraving of a 'Patagonian penguin', trace their ancestry to the start of the Palaeogene, providing clues to the emergence of bird species after the K-Pg mass extinction.

penguins that have ever lived. A little further south, the Antarctic penguin *Palaeeudyptes* reached 2 metres (6½ feet) in height and lived around 35 million years ago. As penguins have denser bones than other birds – an adaptation for life underwater – they tend to survive more often in the fossil record, making them especially useful for understanding the evolution of birds.

Crown of Feathers

Birds are the descendants of dinosaurs, but the origins of modern groups are elusive. Their delicate bones, packed with air to keep them light for flight, mean that their skeletons rarely survive the rough and tumble of fossilization. DNA studies suggest their last common ancestor flapped around in the Cretaceous. The oldest bird branches are the ostriches and kin (such as rheas, kiwis and elephant birds), landfowl and waterfowl. All the other species belong to Neoaves, over half of them being songbirds. These are collectively known as 'crown' birds, indicating that they form the crown of the bird family tree. Their proliferation since the end-Cretaceous mass extinction matches – even surpasses – that of mammals, and they easily outnumber them in almost every habitat.

There are over eleven thousand bird species, from the tiny bee hummingbird (*Mellisuga*) to the hefty ostrich (*Struthio*). They occupy everywhere, including rare microcosms and globe-spanning oceans. The widest wingspan belongs to the albatross (*Diomedea*), a phenomenal 3.7 metres (12 feet) across. Meanwhile, falcons are the fastest vertebrate in the world, reaching speeds of around 200 miles (320 kilometres) per hour plummeting down onto their prey. The most common bird today is the domesticated chicken (*Gallus gallus domesticus*) – we keep around 24 billion of them. Most birds have exceptional eyesight, even seeing into the ultraviolet range. Their respiratory systems are unlike ours, possessing a network of air-filled spaces in the bones. For every inhalation, around a quarter of the air enters the lungs and the rest is sent into their air sacs. Traits such as these give us clues to the biology of their dinosaurian ancestors.

Elephant Bird

Although we call the Cenozoic the 'age of mammals', thanks to the range of body sizes and lifestyles, birds have had their fair share of evolutionary marvels. Among them are the biggest crown birds known to have existed, such as the extinct elephant bird, *Aepyornis*. It resembled a giant ostrich, standing proud at a whopping 3 metres

(10 feet) tall. Despite early travellers' tales of it carrying elephants in its talons, *Aepyornis* was a flightless forager. It probably ate leaves and fruits from the forest floor. Surprisingly, it was not a close relative of ostriches, being more closely related to the kiwi (*Apteryx*) found in New Zealand. Both birds are endemic to their respective islands, meaning that they are found nowhere else. They evolved their unique appearances and lifestyles in isolation. The elephant bird only disappeared in the last few thousand years, coinciding with the arrival of people in Madagascar, who hunted it and ate its watermelon-sized eggs.

Giant birds were not confined to the small islands of the southern hemisphere. In South America during the Palaeogene, there emerged a lineage of monstrous creatures known as 'terror birds', which would have given any Jurassic dinosaur a run for its money. These flightless feathered stalkers ranged from 1 to 3 metres (3 to 10 feet) in height and had powerful legs for running fast. Unlike the tiny-headed *Aepyornis*, terror birds had large skulls with enormous sharp beaks. It is thought they used this to pierce their prey and pull it apart. The larger species became extinct just two million years ago, possibly due to an influx of mammal predators from North America when the two continents were connected together. A few smaller species may have survived to as little as eighteen thousand years ago.

Foraminifera — Telling Environmental Tales

Like miniscule works of art, foraminifera pepper our oceans with their complex shelled bodies. These single-celled organisms are one of the most abundant microfossils in the world. They've survived multiple mass extinctions, and now whisper stories of climate change and environments through deep time.

Foraminifera, or 'forams', are extremely small, shelled organisms, mostly less than 0.1 millimetre in size. Most of them have a shell made up of chambers, either constructed from grains of rock or other shells, or grown by the foram like a snail shell. These shells can be spiral, or have complex symmetries resulting in rice-like tubes, ovals or popcorn-like shapes. Their surfaces are sometimes adorned with spines or ribs, or rough like sandpaper. Although foraminifera are common right back to the origin of complex life, they are especially important for understanding rock layers in the Cenozoic. Many other marine organisms used by scientists for this purpose (such as trilobites and ammonites) became extinct at the end of the Cretaceous, but forams prevail. Their shape, and relationship between different species and their environment, are well-established, making them a superb indicator of environmental and climate change through time.

Forams belong to a group called protozoa, which also includes radiolarians and amoebas. Protozoans are neither animals nor plants, although they often have characteristics we associate with animals, such as the ability to move, and to eat other organisms. Many form partnerships with algae to gain energy from the sun through photosynthesis. The majority of forams, around four thousand species, live on the seafloor (benthic), while a few float in the water column (planktonic). More rarely they are found in freshwater and even in soil. Benthic forams move across the ocean bed using projections from their bodies, called pseudopodia, or cling to surfaces and feed. Others live within the sediment itself, even in the deepest places, such as the Mariana Trench in the Pacific. They are extremely common in tropical regions and near the equator, where fertile marine upwellings stir up nutrients and food. Many feed on drifting food particles in the water, but a few are predators, feeding on fellow forams. Forams provide food for arthropods, fish and birds, making them a crucial part of ocean food webs. Despite their tiny

Fossils of foraminifera, pictured here with other small organisms, can be studied to reveal past climates and ocean environments.

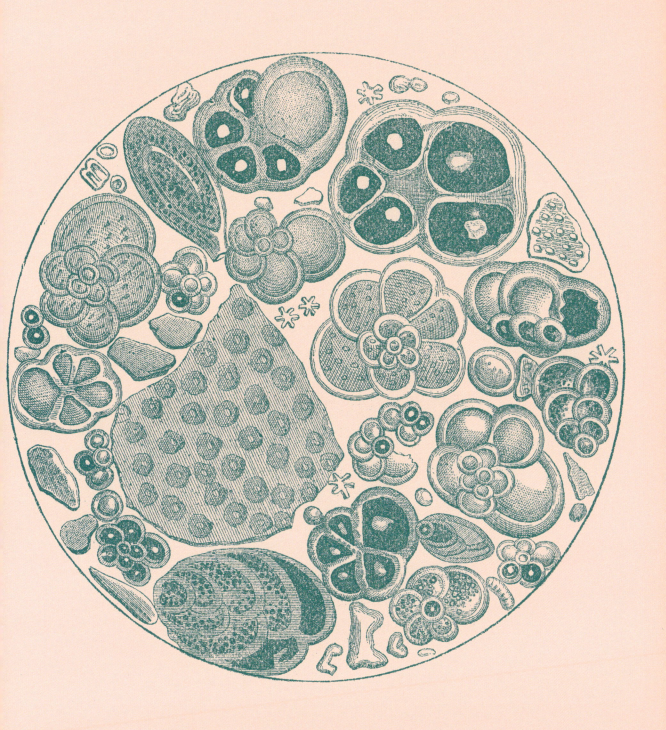

size, these little organisms also keep a record of the conditions on our planet that would otherwise be unknown to us.

Telling Environmental Tales

Thanks to the stories these tiny creatures contain about the past, foraminifera have proven invaluable for reconstructing climate in deep time. Different species tell us what the ocean environment was like, and their fossils can be analysed for isotopes and trace elements, which change depending on the planet's carbon cycle, temperature and continental weathering. They are especially sensitive to ocean acidification and to general climate changes. Deep-sea drilling programmes have brought up thousands of cores from the seabed to examine the fossil forams and pinpoint oil and gas reserves. This has resulted in a comprehensive global record that stretches back millions of years.

As they build their shells from minerals in seawater, fossil foraminifera are like time capsules filled with the 'flavour' of the sea at different points in the past. For example, when tectonic plates lift to form mountain chains, these peaks are worn down by rain and the chemicals washed into the ocean, where they are incorporated into foram shells. As a result, forams provide evidence for changes taking place on a global scale – and on land, not only in the ocean. They have even been used by archaeologists to trace materials used by ancient peoples back to their source, matching the forams in the rocks to their origin.

Thermal Maximum

Until recently, our planet was especially cool – in fact we are currently living in an interglacial period, merely a resting point between ice ages. During the Palaeogene, the Earth was far warmer than it is today, especially 66 to 34 million years ago. The Palaeocene–Eocene Thermal Maximum (PETM), marked a peak in global temperature 55 million years ago, when it was so warm that there were crocodiles and palm trees in Antarctica. The heating is thought to have been caused by volcanic eruptions in the North Atlantic, triggered by the splitting tectonic plates between Europe and North America. Evidence of these volcanos can be found dotted along the shores of Ireland, Scotland, the Faroe Islands and Norway; the Giant's Causeway in Ireland, for example, is a shoreline of rock pillars formed from cooling lava. These eruptions poured thousands of gigatons of carbon dioxide into the atmosphere, causing massive global warming.

This probably triggered feedback loops, releasing methane from the deep sea and raising temperatures even further.

The PETM is incredibly important as a model for understanding the effects of human-induced climate change. The foram fossil record tells us that during the PETM, even the deep ocean acidified, and in as little as a thousand years, 50 per cent of benthic forams became extinct. Temperatures rose by about 6 degrees Celsius (10.8 degrees Fahrenheit) in twenty thousand years, a startlingly rapid increase. Although global temperatures fluctuate naturally, abrupt changes usually have devastating impacts on animal and plant life because they have little time to adapt. If our current rate of human-induced climate change continues, it will be one hundred times faster than the PETM, with some estimates predicting a 6-degree Celsius (10.8-degree Fahrenheit) rise as early as the year 2100. With so little time to respond, life on Earth is facing an extinction event far greater than the PETM, or perhaps any other extinction event in history.

Ants — The Social Insect

Perhaps the most astonishing accomplishments of any animal on our planet belong to the industrious ant. They build metropolises, shape habitats and exhibit an astonishing array of intimate relationships with other insects and plants. In the Palaeogene they became a cornerstone of life on Earth, now comprising up to a quarter of the biomass in some tropical habitats. Their study has transformed our understanding of natural selection.

Ants can be found on almost every continent and most islands. Their tiny armies sweep through forests, invade homes and build fortresses to rival any human capital. Although individually small, their boggling numbers mean they often outweigh the animals they live alongside: as much as a quarter of animal biomass in an ecosystem comprises ants, particularly in rainforests. They aerate soils and recycle nutrients almost as effectively as earthworms. They are prolific predators and substantial herbivores. Their social lifestyle not only makes them adaptable to most climates, but also allows them to alter their surroundings. The major groups of ants we know today had all appeared by the Palaeogene. It was in this time period that they took on a pivotal position in our planet's ecosystems.

Ants are related to wasps and bees, and although often confused with termites, they belong to completely separate branches of insects. There are over thirteen thousand different species of ant on Earth today, including species smaller than a quinoa grain. The ant fossil record is extremely rich, and although the oldest ones are known from Cretaceous amber, their origins likely lie in the Jurassic. In the Palaeogene, warm temperatures created giants such as *Titanomyrma*, which lived in North America and Europe and was the size of a hummingbird.

Ants have distinct bent antennae that detect chemical scents, air currents and vibrations. Some have exceptional eyesight, while a few underground specialists are completely blind. Two strong jaws are used for carrying, constructing and fighting, and most ants are exceptionally strong for their size. Their co-operative colonies vary from a handful of individuals to enormous metropolises of millions. Larger colonies usually have strict castes comprising workers, soldiers and one or more fertile queens – some queens survive up to thirty years. These communities are so integrated they can be

There are over thirteen thousand species of ant, including leafcutter ants like *Atta cephalotes* (top middle), and the honeypot ant *Camponotus inflatus* (bulbous ant, bottom left).

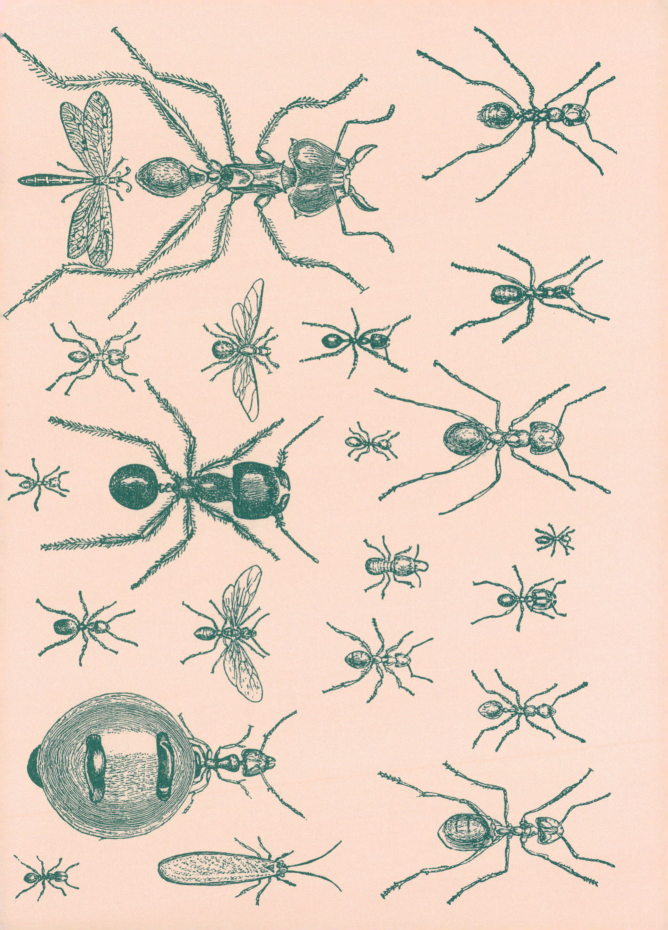

considered superorganisms in their own right. Ant biology and behaviour has been crucial for our understanding of evolution, particularly kin selection and co-operation. Since their appearance, ants have become one of the dominant insect predators and scavengers, and have developed thousands of surprising relationships with other organisms, including plants, fungi and micro-organisms.

Complex Societies

We think of ants as colonial nest builders, but not all of them live the same way. Parasitic ants enter the nests of their hosts and exploit them for resources, whereas slave-raiding species capture the worker eggs or larvae of other ants and incorporate them into their own colony. Perhaps the most famous are the army ants, which barrel across habitats like a tsunami, attacking anything in their path – including humans.

The division of labour in ant colonies allows them to solve problems insurmountable to lone insects. They can remove floodwater from nests by drinking it and excreting it outside, or form living rafts to float to safety. Others use their own bodies to bridge gaps in difficult terrain for their comrades to cross. They can forage over 200 metres (220 yards) from their nest, using pheromone scent trails to lead them home. Fellow workers follow these trails to relocate resources, with the sun's position and Earth's magnetic field as additional guides. Ants communicate by scent, touch and sound, and often release an alarm pheromone when attacked, driving the nest into a defensive frenzy. Many spray chemicals or sting – the bullet ant (*Paraponera*) is said to have the most painful sting of any insect. Despite this, mammals such as anteaters, echidnas, pangolins and numbats have evolved specialisms for ant-eating, such as long sticky tongues and strong forelimbs and claws to break into nests.

First Farmers

Humans think of themselves as the ones who developed farming, but ants beat us to it by 66 million years or more. A few ant species farm other insects for their honeydew: for example, plant-eating aphids, which secrete liquids packed with sugar. This manna is drunk by the ants, which protect their 'herds' from predators and tend to them, like shepherds to their flocks. They may even take their aphids with them when they move the nest. Similarly, some caterpillars are raised by ants, who put them out to eat their favourite plants during the day, returning them to the safety of the ant nest at night. Such behaviour could have provided a path for the evolution of sociality in ants, who had to work together to protect their food source.

Other species, such as leafcutter ants, cut and collect leaves, bringing them back to a 'garden' in the colony. There, they snip them into very small pieces and use them to raise fungi, which are their main food source. Like human agriculturalists, they tend their gardens, removing anything toxic or harmful to the crop. They even harbour bacteria on the external surface of their bodies that produce antibiotics that kill micro-organisms harmful to the fungi. Both the ants and fungi need one another to survive – in many cases the fungi can no longer grow outside of ant-farmer gardens, essentially making them a domesticated species. The leaf-gathering of leafcutter colonies can account for as much as 15 per cent of animal herbivory in an ecosystem.

There are plants that rely on ants to disperse their seeds, or to protect them from animals. The bullhorn acacia of Central America has hollow thorns that host ant colonies, protecting the tree from parasitic vines and browsing mammals in exchange for shelter and food. Lemon ants will kill all the other plants surrounding their favoured tree, the lemon ant tree (*Duroia*), where they like to build their nests. This kind of behaviour shapes the landscape itself.

Neogene

The Neogene spans just over twenty million years, from the end of the Palaeogene to the start of the Quaternary. The rise of monumental mountain chains redirected Earth's climate, cooling it and aiding the spread of grasslands. The first horses galloped over the steppe, while beneath the waves kelp forests expanded, sheltering resources that sustained the first humans foraging along the shore.

The Neogene began 23 million years ago and ended just 2.6 million years ago. The continents reached the positions we recognize, although their outlines were not always the same thanks to changing sea levels, at times 20 metres (65 feet) higher than today. Until the very end of the period, the climate was warmer than today, although it had been cooling steadily since the hothouse of the Palaeogene. Earth was still not quite the world we know. The polar icecaps had just begun to form, and as they did, sea levels dropped, exposing bridges of land between once isolated continents. This wreaked havoc, as animals migrated to new territory and competed with the inhabitants. The Tethys Sea finally closed, joining Africa to Europe and forming the Mediterranean Sea. Near the end of the Neogene, this body of water dried up multiple times, as the ice ages caused sea levels to plummet.

The climate and environments of Earth changed radically in the Neogene. Deserts expanded across central Asia, the Sahara and parts of South America, and Australia desiccated as rainfall lessened. In this drying, cooling world, tropical forests shrank, while grassland – a small component of our planet's flora until this point – took root in their wake. Enormous savannahs developed in tandem with the animals that grazed them, including horses, antelope and elephants. Cats, dogs and kin took over as the main carnivores on land, while some of the largest sharks to have ever lived, such as *Megalodon*, shared their watery world with new species of whales and seals. The first kelp forests evolved in our seas, creating the most productive habitats on the planet.

Great American Interchange

For around 100 million years, South America had been isolated. It lay on the other side of the widening Atlantic from Africa, and cut off from North America by an equatorial seaway called the Strait of Panama. This kept its animal inhabitants separated from those of its neighbours, resulting in the evolution of unique plants and animals, including marsupials. Around three million years ago, at the end of the Neogene, the Strait of Panama closed, joining North and South America for the first time since the Cretaceous.

Animals and plants began to cross between the continents, an event known as the Great American Biotic Interchange (GABI). The GABI is one of the most important topics in palaeontology and ecology because it allows a rare, detailed glimpse at the effects of isolation and introduction of animal groups. It was first discussed by the co-discoverer of the process of evolution, Alfred Russel Wallace (1823–1913), who spent time in the Amazon Basin. Hoofed mammals (horses, tapirs and camels) as well as cats, dogs and bears travelled down into South America. Meanwhile, southern animals spread north, including capybaras and armadillos, and many now-extinct creatures such as ground sloths and terror birds.

Although the exchange was more or less even in both directions, in the long run the North American animals fared better than their southern counterparts. It is thought that a combination of direct competition and climate changes allowed the northern species to survive. For animals moving south, the habitat was less varied, probably proving less of a challenge than to those moving in the opposite direction. Many questions remain unanswered, but the GABI resulted in the patterns of species we see in the Americas today.

Everything is Connected

Although we know habitats have changed over millions of years, the picture is more detailed for more recent time periods. In the Neogene researchers can tie the appearance and expansion of new ecosystems, such as grassland and kelp forests, to events taking place on a global scale. In the Neogene there are some incredible examples of how plate tectonics

affect the whole world. The formation of the Isthmus of Panama, connecting North and South America, may have provided a bridge between those land masses, but it also erected a permanent barrier between oceans. Warm currents from the Pacific no longer flowed into the Atlantic, contributing to the onset of ice ages. Similarly, as the Indian subcontinent pushed northwards into Asia, it shifted the ocean and atmospheric currents of our planet, triggering a new climate cycle called a monsoon. From July to September, moisture-laden clouds are pulled in from the Arabian Sea and Bay of Bengal, and move northwards across the land, as far as Tibet and China. The Himalayas block their passage and force the clouds upwards, causing them to release deluges of rain. The monsoon accounts for around 80 per cent of rainfall in India, upon which much of the country's agriculture depends.

The rising Himalayas and monsoons had an altogether unexpected repercussion for other parts of the world. As the water ran over the rising mountains, it eroded the rocks, drawing down carbon dioxide from the atmosphere in a process known as silicate weathering. This reduced CO_2 levels in the atmosphere, cooling the planet further. A similar process was happening in South America as the Andes mountain chain was pushed up along the western coast. The subsequent ice ages that dominated life from the end of the Neogene and into the Quaternary are thought to have been triggered by these global changes.

Grasses — Shaping Animal Life

Our civilizations are built on grasses, feeding billions of people around the world. Grasses grow on every continent, coating up to two-fifths of the Earth in an emerald pelage. These blades formed the first savannahs in the Neogene, literally altering the shape of animals that relied on them. From our beloved front lawns to massive monocultural harvests, our relationship with grasses has not only shaped our past, but also plays an important part in a sustainable future.

Grass is so everyday we tend to overlook it, but these flowering plants cover up to 40 per cent of our planet. From pampas to prairie, steppe to savannah, grass has played a fundamental role in the formation of modern ecosystems. In the Neogene it spread as the climate cooled, dominating huge portions of the world for the first time and shaping the animals that depended on it. Despite this, because its fossil record is often limited to microscopic structures such as pollen, there remain many unanswered questions about the origins and evolution of this most glorious of greens.

Grass grows in a distinctive shape, with an upright hollow stem and long tough leaves that are flat and pointed. The flowers form a spikelet and are wind-pollinated, one of the leading causes of hay fever in humans, an allergic reaction to plant pollen. Grass grows everywhere – even in Greenland and Antarctica. Antarctic hair grass (*Deschampsia*) not only braves the extreme winter conditions, but is also extending its range poleward as our world warms through human-induced climate change. Looking back through deep time, for much of our planet's history grass did not exist. The oldest-known fossil grass dates to the Cretaceous, when it emerged alongside other flowering plants as part of the Cretaceous Terrestrial Revolution. Early grasses were probably not especially common, likely growing in forest margins or shade – some grasses today, such as bamboo, still enjoy those conditions. What made them so successful in the Neogene was their startling adaptation to drier, open habitats, and their tolerance for drought.

There are around twelve thousand species of grass, making it the fifth-largest plant family. Staple human foods around the world come from this generous group, providing over half of all the energy we consume. They are fed to livestock; used for construction, such as bamboo, straw and thatch; or used as fuels, from fire-lighting to biofuel. Yet grasses have played a much more important role in

Five meadow grasses including cock's foot grass (*Dactylis*), fescue (*Festuca*) and dog's tail grass (*Cynosurus*). Their successful spread is linked to global climate change, and the emergence of many mammal herbivores – as well as human societies.

shaping the bodies of other animals. The intimate tie between them and grazing mammals is clear from the physical adaptations each has undergone. Grass was a game-changer for our planet, and those that walked its many continents.

Fuelling the Herds

Since their emergence in the Cretaceous, grasses and herbivorous animals have relied on one another. Trees and other plants outcompete grasses over time, but grazing animals trample and eat this competition. Grass grows from its base, easily surviving herds of hungry mouths, wildfires and mowing by humans. We know sauropod dinosaurs fed on grass in the Cretaceous, thanks to the contents of their fossilized dung, which contains microscopic structures in grass called phytoliths. These are made of silica, and some are so sharp they can cut human skin. To counteract such damage, grazing mammals have evolved longer teeth with extensive enamel and increasingly complex furrows, making them more resilient. Such teeth are found in cows, horses, elephants, rabbits and rodents. Elephants and grazing marsupials also uniquely replace their molar teeth periodically in adulthood.

The grassland environment has also had dramatic effects on the shape of mammals. Hoofed mammals such as horses and deer have evolved longer legs with reduced numbers of digits (fingers and toes), and the joints of the leg move front to back, as opposed to swinging out to the side of the body. These adaptations improve the efficiency of movement, making it easier to cover long distances and run fast. As herds move seasonally across grasslands they are more vulnerable to predators, so speed and endurance are vital for their survival.

Human Staple

Our main grain and cereal crops are grasses. The oldest evidence of humans eating them dates to around 105,000 years ago in Mozambique, but around the world we've been growing wheat, rice and maize since at least 11,500 years ago. There is evidence for rice cultivation in China 7,700 years ago, in the coastal wetlands near modern-day Hangzhou, while at the same time in Mexico, maize was domesticated from wild teosinte. The domestication of such grains has increased their yields, producing far larger quantities than the wild ancestral plants ever could have.

Intensive, monoculture farming has a radical negative impact on our natural world. It has resulted in loss of habitat and biodiversity, while pesticides and fertilizers damage wildlife and waterways, decimating insect life. Agriculture consumes around 70 per cent of our freshwater worldwide. It takes 1,000 litres of water to produce just 1 kilogram of grain, a resource usage compounded if that grain is used to feed cattle (43,000 litres produces just 1 kilogram of beef). Water shortages already exist around the world, with over a billion people having no access to adequate drinking water. With further water shortages predicted in the future, efforts are underway to make farming more water-efficient, and breed – or even engineer – grasses and other plants that can be farmed in extreme water scarcity.

Along with food, the modern lawn has become an obsession, whether it is a golf course or the patch in front of your house. In parts of the world unsuitable for such turf, huge quantities of water are used to maintain it, contributing to drought. Our use of grasses for human food is a success story for this plant, which has long exploited mammals to spread around the world. But the drawbacks to this bountiful resource are being felt, prompting a rethink on the role of this wonderful, world-shaping plant in a sustainable future.

Merychippus — Evolution of Horses

Merychippus was the world's first recognizable horse. The transformation of horses from small dog-sized ancestors to the magnificent animals that powered human expansion around the world is a textbook example of the ties between animal anatomy and habitat. But the bushy branches of their family tree also remind us there is no defined end goal to the merry gallop of natural selection.

In the Neogene, evolution assembled the horse as we know it. *Merychippus* was not much larger than a Shetland pony. It lived between 16 and 5 million years ago in North America, just as grassland habitats were replacing forests across huge tracts of the landscape. At only a metre (3 feet) at the shoulder, for much of that time it was the tallest horse in existence. It had characteristics we recognize in horses today: a distinctive long face with ears high on the head, a long neck and slender legs. *Merychippus* is significant because it was one of the first horses fully adapted to grazing on grass. It had wide molar teeth with plenty of surface area and lots of enamel (known as hypsodont teeth) able to withstand the hard wear imposed by silica-rich grass blades. Although it still had noticeable second and fourth toes, the hoofed middle toe on each leg bore its weight, and was supported by strong ligaments, making it ideal for running across open plains.

Horses are hoofed mammals, and alongside zebras, rhinoceros and tapirs, they belong to a group called odd-toed ungulates, or perissodactyls. All of them support their weight through their third, middle toe. In the case of horses, they have lost almost all trace of the other digits. The counterparts to this group are the even-toed ungulates, or artiodactyls, which include deer, pigs, giraffes, camels, llamas, sheep and cattle. They support their weight between the second and third toes instead. The loss of toes and elongation of their legs were an adaptation to the expanding dry grasslands of the Neogene.

Modern horses originated in North America in the last million years – relatively recently in geological terms. The largest wild horse is the Grévy's zebra (*Equus grevyi*), at around 1.4 metres (4½ feet) tall, but domesticated working breeds of horse, such as the Clydesdale, can stand over 1.9 metres (6 feet) tall. Horses all have manes and long hair on their tail. Zebras have the most extravagant fur patterns; their

The skull of *Merychippus*, with a characteristically 'horsey' long face and wide molar teeth perfect for grinding grass.

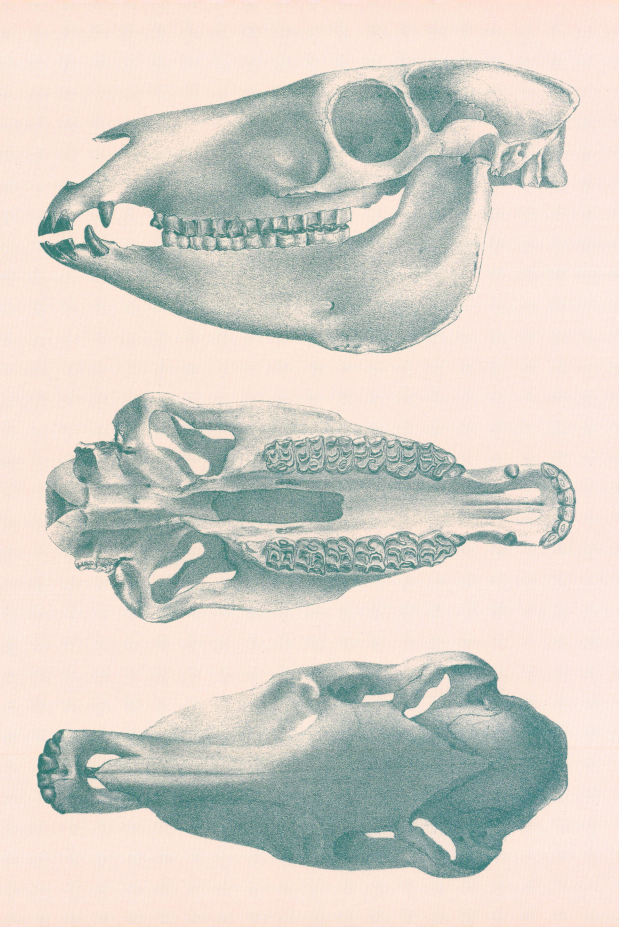

bold black-and-white stripes deter biting insects and confuse predators. Most horses live in social groups, often family harems with a single male alongside females and their offspring. Horse evolution is especially significant for humans, who have relied on their muscle for as much as five thousand years, and still use them as a measure of power for motor engines today.

Progressive Evolution

The evolution of horses is often seen as a straight line, from compact forest dwellers to the tall, powerful animals we know today. Yet their emergence was not linear from ancestor to modern breeds; there were many different kinds of horse at different points in their evolutionary history, most of them later dying out. The origins of the group lie with creatures such as *Sifrhippus*, which lived around 55 million years ago in North America. It was fox-sized, with short legs and five-toed feet. Today's horses all belong to the genus *Equus*, which includes horses, donkeys, onagers, kiangs and zebras. Studies of DNA suggest all of them can be traced back to a common ancestor at the end of the Neogene, around five million years ago.

When Europeans reached the Americas there were no wild horses to be found there, but, surprisingly, horses originally emerged on that northern continent. Clues to their origins first came from fossil horses found in the Americas in the 1700s. During his time sailing around the world on the *Beagle*, Charles Darwin was shocked to find fossil horse teeth in Patagonia, alongside extinct giant armadillos. Hunting by humans may have contributed to their extinction in the Americas, alongside climatic changes during the ice ages. As more fossil horses were discovered in the 1800s, their story came to represent evolutionary progression: the notion that evolution progresses in a straight line from one form to the next towards a clear end goal. We now know that this is not the case for any animal, and instead there are multiple short branches of animals that become extinct, with no end goal, simply adaptations to ever-changing ecological conditions.

Domestication of Horses

If there is one mammal that has transformed the course of human civilization, it is the horse. Humans have been fascinated by them for thousands of years, painting them as cave art as long as thirty thousand years ago, and hunting them for meat and skins. The oldest evidence for horse domestication comes from Kazakhstan, where they played a central role in Botai culture over five thousand years ago. As well as riding horses, traces of horse milk in pottery shows that they were farming these animals, which were probably domesticated from wild Eurasian herds.

By around four thousand years ago, humans had begun using horses to pull chariots. After that, domesticated horses spread rapidly across Europe, northern Africa and China, being ridden and pulling vehicles for warfare, farming and construction. For thousands of years, they provided the primary form of power (along with cattle), able to pull twice their body weight and carry up to 100 kilograms (220 pounds). They allowed people to cover huge distances quickly, as much as 100 miles (160 kilometres) in a single day, and can also reach speeds of around 35 miles (56 kilometres) per hour over short stretches.

The ancestors of the modern horse were the only survivors of their lineage. Although 'wild' horses such as mustang live in places such as Central Asia, Australia and the Americas today, we know from DNA analysis that all of them come from feral domesticated horses. Przewalski's horse (*Equus ferus przewalskii*), found in Asia, was thought to be a true wild horse, but similarities between its genes and those of domesticated horses in archaeological excavations of the Botai culture suggests that it is the descendant of escaped domesticated animals, rather than a remnant of truly wild animals.

Tuatara – Unique Survivors

The tuatara is the last survivor of a once globally common reptile group. This knobbly noble is one of many rare and unique organisms in New Zealand: the Earth's southern time capsule. These incredible species teach us about the distribution of life across continents, and their biology provides glimpses into the ancient patterns of evolution. Despite enduring millennia of geological and climate change, invasive species and habitat loss now threaten them with extinction.

The tuatara (*Sphenodon punctatus*) is a reptile found only in New Zealand. It looks like a lizard, with limbs sprawled to the side and dun-grey to yellow scaled skin. It can grow to around 80 centimetres (31 inches) long, with a row of small spikes protruding along its back like a picket fence – in the Māori language its name means 'spiny back'. Although it has ears, there are no earholes, and its eyes are large and appear almost black. It may be mistaken for a lizard, but this unique creature is the last living representative of an entirely seperate group of reptiles that was once among the most successful on our planet.

The tuatara is a rhynchocephalian (*rinko-seph-alien*), a reptile group that shares a common ancestor with squamates (lizards and snakes). But these two lineages split apart over 240 million years ago in the Triassic. The oldest fossil rhynchocephalians are found in Germany, and they once lived across most of the Pangaean supercontinent. Throughout the Mesozoic they were diverse and widespread, similar to lizards today, with meat-eaters, plant specialists and shell-crushers, aquatic species and snake-like forms. However, in the Early Cretaceous rhynchocephalians begin to disappear, and by the earliest Palaeogene had gone from everywhere except New Zealand. It is uncertain what drove them to extinction elsewhere; perhaps competition with squamates and predation by new species of mammals and birds contributed to their decline.

Although there are now over 10,600 species of lizards and snakes, there is just a single rhynchocephalian, the tuatara. They are nocturnal, and eat small vertebrate animals and eggs. They sun-bask during the day to warm their bodies, but unlike lizards, can function in colder temperatures, growing and reproducing very slowly. They are long-lived, reaching sixty years old in the wild, and able to reach over a hundred years in captivity.

The tuatara may look lizard-like, but is the last of an ancient, unique lineage of reptiles that once inhabited the whole globe.

216

Last of their Line

The tuatara, and sole survivors like them, teach biologists about rates of evolution and common ancestry among living animal groups. A similar example are the platypus and echidnas, the only living members of the mammal branch called the monotremes. Like the tuatara, they are restricted to a small region of the world – Australia, Tasmania and New Guinea – despite their ancestors having lived all across the northern continents and South America. Monotremes have unique features not present in other mammals, including laying eggs.

Although animals such as these are sometimes called 'living fossils', this term is not meaningful. They may be the last of their kind, but their biology and DNA reveal they have undergone a great deal of molecular and anatomical evolution, even if their superficial appearance looks unchanged. There are many possible reasons why such animals survived when the rest of their family has disappeared. People have assumed that their isolation protected them while their kin were out-competed by new species, but this is probably too simplistic. Monotremes, for example, continue to thrive alongside both marsupial and placental mammals in Australia, and the tuatara co-exists with lizards in New Zealand. It is more likely a complex interplay of changing climate, habitat and isolation – as well as plain old luck – that has resulted in the patterns of life we see on our planet today.

Island Isolation

New Zealand differs from other parts of the world thanks to its geological history, which has kept it separated for over seventy million years. It belongs to a chunk of continental crust called Zealandia, or Te Riu-a-Māui, which was once part of the great southern continent of Gondwana. Around eighty million years ago, it broke away to become its own microcontinent. This microcontinent was often partly submerged beneath the sea, but around forty million years ago volcanos erupted, creating new land. In the Neogene, a fault line lifted the rocks to form the Alps of New Zealand's South Island. Recent ice ages lowered sea levels, further exposing the edges to form the coastline with which we are now familiar.

New Zealand is one of the only places where we still find organisms once common across Earth's southern continents. They make it an important biodiversity hotspot. There are trees such as the kauri (*Agathis*) and the southern beech (*Nothofagus*) – some of these and their close relatives are also found in South America and Australasia. Birds such as the kiwi and kakapo, and the extinct giant ostrich-like moa, became flightless in the absence of large ground predators. One of the only native mammals is the short-tailed bat (*Mystacina*), which forages on the forest floor, making it the most terrestrial bat in the world. Many of these organisms share common ancestors with relatives in South America and Australia, suggesting they may have arrived in New Zealand from those continents over the last thirty million years.

Invasive Species

Like many of New Zealand's unique and wonderful wildlife, the tuatara is under threat from invasive species. These are organisms introduced to habitats where they were not previously present, with severe negative consequences. Although invasive introductions happen naturally when barriers are removed, the majority of invasive species today are the result of human introductions. Although humans have been dispersing species around the world for millennia, this has increased rapidly since the 1700s, as European colonization began and international trade routes expanded.

Among the best known invasive species is the Polynesian rat (*Rattus exulans*). Originally from Southeast Asia, it was probably a stowaway on boats travelling between the Pacific Islands over the last two thousand years. Multiple Pacific Island bird and insect species have become extinct as a result, including in New Zealand. They may even have contributed to the deforestation of Easter Island, by eating palm-tree nuts, so preventing their regeneration. The roots of Japanese knotweed (*Reynoutria japonica*), introduced to Europe and North America, damage building foundations and roads, as well as crowding out native plants, making it one of the world's worst invasive species.

The tuatara faces predation by non-native predators such as cats and rats. As a result of their introduction, it became extinct on mainland New Zealand, surviving only on small offshore islands. It has now been reintroduced into a sanctuary on the mainland North Island, and is breeding successfully in the wild again. But it, and the many other unique native animals in New Zealand, face an uncertain future. Habitat loss and climate change add to their already precarious status. The loss of the tuatara would not only be a tragedy for biodiversity, but would also be the end to an incredible and ancient lineage of truly unique reptiles.

Kelps – Most Productive Ecosystem

Kelps are the rainforests of the ocean, from their dark holdfast trunks to the dappled sway of their canopy. All seaweeds are algae, one of the first groups of organisms on the planet, but kelp forests only appeared in temperate seas as the planet cooled in the Neogene. They are home to a stunning array of marine life, providing sustenance to humans for thousands of years. Kelp forests are also indispensable to our understanding of food webs and the repercussions of unbalancing nature's intricate ecology.

Kelps – for there are more than one of these wondrous species – are some of the most significant organisms in the world. Yet we may live our whole lives entirely unaware of their fundamental role in the ocean. They are the rainforests of the sea, covering thousands of square kilometres of the seabed. Generally brown in colour, kelps are a type of seaweed found around the world on temperate and polar coastlines. They are made up of fronds, or 'blades', which grow at the phenomenal rate of up to half a metre (20 inches) per day, reaching as long as 60 metres (66 yards). At their base they grip the seafloor using a root-like structure called a holdfast. Although in some ways they resemble plants, seaweeds are actually algae, one of the oldest organisms on Earth, with over a billion years under their belt – twice as long as plants and animals. The modern kelps that make our waters so fecund owe their origins to the cooling climate of the Neogene.

Algae are photosynthesizing organisms. They can be single-celled, such as diatoms, or form complex multicelled structures, such as seaweed. Although they also use sunshine to generate energy, seaweeds have a completely different structure from plants, and can live in fresh or salt water. There are around 120 species of kelps, and they vary enormously in shape and size. Some have gas-filled bladders that keep their fronds afloat, while others lie flat along the seabed. Kelp forests are dense and often thrive when upwelling currents from the deep deliver cold, nutrient-rich water to mix with surface currents. Like a forest on land, they have a dense canopy, producing a shadowy micro-environment nearer the seafloor.

Kelp forests provide habitats for thousands of other organisms: as many as a hundred thousand invertebrates can inhabit a single square metre. Kelp forests are home to shrimp, snails, bristle worms and urchins, and support myriad shoals of fish and marine mammals and birds, including terns and cormorants. Altogether, they are one

These kelps belong to the order Laminariales, a brown algae that often forms underwater 'forests', home to thousands of species of animal.

of the most productive ecosystems on our planet, not only for wildlife, but also for humans too. Since the first people began exploring Earth's temperate shorelines, kelps have provided food, materials for crafts and construction, and, latterly, chemicals for industry. As we destabilize our planet's climate and food webs, these startlingly bio-diverse habitats are under threat, risking a loss that would ripple through every part of our world.

Cooling Ocean Habitats

For a long time, the origins of kelps have been unclear, but we now know that Earth's cooling climate over the last thirty million years led to their expansion. It is thought they first evolved in the northern Pacific, which may explain why the coasts of Japan and North America hold the richest abundance of species.

The evolution of kelps is also linked to many other animal groups. We know kelp forests had emerged by the end of the Neogene thanks to the fossils of limpet species that feed on them, which don't appear until that time. At the other end of the food web, sea otters (*Enhydra lutris*) have adapted to inhabit kelp forests. These special-ized animals are mustelids, a type of carnivoran mammal related to weasels and badgers. Although they are the largest of their kind, at an impressive 45 kilograms (100 pounds), sea otters are the smallest marine mammal in the world. They are so comfortable in the water that they sleep floating on the surface, wrapping kelp strands around their bodies to prevent from drifting away on the current. They're also one of the few animals that use tools; employing rocks to smash open hard seashells.

Kelp Industry

The emergence of kelp forests has not only effected ocean biodiver-sity, but also human evolution. Ancient Stone Age settlements often include evidence that people were eating the animals that live in kelp forests, such as abalone and limpets. It has even been suggested that their rich natural resources made it possible for humans to disperse around northeast Asia into the Americas, using it like a 'kelp highway'. Kelps such as the bull kelp (*Nereocystis luetkeana*) were used to create fishing nets, and around the world, coastal communities enrich soil for farming by adding kelps as fertilizer.

Seaweeds such as kelps are packed with iodine and alkali used in human industrial processes. In the 19th century they were har-vested and burned to produce soda ash, used in soap and glass production. In the Scottish Highlands, demand for soda ash led land-lords to force tenants to harvest seaweeds, depriving them of the ability to make a living by any other means. The enormous profits of this industry were not passed to the impoverished tenants, and this was a huge contributor to the Highland Clearances, which saw Scots emigrate to colonies around the world. Kelp extracts can also be used as a thickener for foods such as jelly and toothpaste, as well as being eaten. Kombu, for example, is an important ingredient in Asian cooking. As it grows quickly and can be harvested from the surface by boat, kelp is a particularly productive and easy food to grow. It is also extremely environmentally friendly. Research is currently underway into using kelp to produce biofuels.

Urchin Barrens

Kelp forests have provided ecologists with a foundational example for understanding trophic processes: how living things interact with the rest of their food web. For thousands of years, humans hunted sea otters along the North American Pacific coast. When colonists arrived from around the world in the 18th century, one of the resources they exploited was sea otter fur; the densest in the world and sought-after for clothing and accessories. Well over a million sea otters were slaughtered, wiping them out entirely across much of their range.

The loss of sea otters caused a 'trophic cascade'. This occurs when part of a food web is radically reduced or removed, disrupting the balance of the entire ecosystem. Sea otters are a primary predator of sea urchins, a round creature covered in a spiky hard shell. With sea otters gone, urchin numbers exploded. The urchins feed on kelps, and their unchecked armies destroyed hundreds of square kilo-metres of kelp forest, creating 'urchin barrens' where very little could survive. This remains one of the clearest examples of the crucial role of predators – the 'top tier' of food webs – in the health of the whole ecosystem, and the far-reaching consequences of losing even one species from the tapestry of life.

Humans have benefitted throughout history from kelps and the animals that rely on them, but thanks to our activities this beautiful ecosystem is under threat. Pollution and climate change have had a drastic effect on kelp forests, as have invasive species. Coupled with overfishing and hunting, the impact has been disastrous along many coastlines. Unless we act fast, we may lose these incredible havens of life forever.

Quaternary

The Quaternary commenced 2.6 million years ago in the icy grip of glaciation. As the deep freeze spread across the highest latitudes, it carved out the landscapes we know. Humans have colonized the world in the last slice of this brief period of time, wiping out hundreds of species in our wake. We are the only technological animal, the first organism to consciously hold the future of life in our hands.

The Quaternary is the shortest geological period since life began on Earth – but admittedly, it is a work in progress. It is sub-divided into the Pleistocene and the last 11,700 years, called the Holocene. Only in this last sliver of time have complex human societies arisen.

Although our planet's continental configuration has changed little in the Quaternary, cycles of ice ages have periodically drawn up massive volumes of fresh water, lowering global sea levels by over 100 metres (330 feet) and revealing land bridges between continents. The Great Lakes in North America, which now hold 21 per cent of the world's fresh water, were carved out by moving ice sheets, then deepened and filled as the glaciers melted. Other parts of the world dried out, extending the arid regions that first appeared in the Neogene to form the Sahara, Namib and Kalahari deserts.

Most Quaternary animals are easily recognizable, but there are some exceptions, such as giant ground sloths and strange giraffe relatives. Ice Age mammals such as sabre-toothed cats and woolly mammoth flourished at the fringes of the glaciers. Many became extinct around 11,500 years ago. These disappearances often coincide with the spread of modern humans, but we still don't know to what extent our appearance contributed to their demise. Other Quaternary extinctions include the giant elephant bird of Madagascar, moa in New Zealand and, more recently, many hundreds of animals including the dodo, thylacine and passenger pigeon. Human-caused habitat destruction, pollution and climate change are now the greatest threats to biodiversity on our planet.

Ice Ages

Although ice has always been a fluctuating feature of planet Earth, the Quaternary has seen the most widespread and consistent ice cover since 'Snowball Earth' at the end of the Proterozoic. At their zenith glaciers clasped our globe from the poles as far as 40 degrees latitude, with permafrost reaching even further. The ice sheets have fluctuated in size and extent, with their maximums known as glacials, and their minimums called interglacials.

Around twelve thousand years ago, at the last glacial maximum, nearly a third of the Earth's surface was covered by ice, including Europe, Russia, Mongolia, northern China, Alaska and Canada. In the south, glaciers covered Patagonia, and New Zealand. These regions are now experiencing 'post-glacial rebound', where the tectonic plates that were pushed downwards by heavy ice sheets spring upwards again, causing the land to rise by an average of 1 centimetre per year, often faster.

Many landscapes in the far northern and southern hemispheres owe their shape to the ice ages. The glaciers were as much as 3 kilometres (nearly 2 miles) deep, and although they appear static, ice actually moves, like a slow-motion river. It scrapes the land beneath, removing truckloads of soil and rocks, and carving out distinctive U-shaped valleys.

Milankovitch Cycles

To understand how our climate works, we must look not only at our atmosphere and water cycle, but also at the movements and tilt of the Earth. Our planet doesn't sit upright in space, but tilts and wobbles as it spins on its axis around the sun. The patterns of these movements were calculated by Serbian astronomer and geophysicist Mulitin Milanković (1879–1958), after whom they have been named.

There are three main Milankovitch cycles. The first is eccentricity, which describes how the Earth moves around the sun. Over the course of 413,000 years, this circuit varies from circular to an oval, or elliptical, orbit. The second cycle is tilt, the degree to which the poles are tilted from vertical, which varies between 22.1 and 24.5 degrees every 41,000 years. The final cycle is precession, the 'wobble' of the planet, which occurs over 25,771 years,

making it the shortest cycle of the three.

These cycles have a profound impact on our climate because they alter how close we are to the sun, how much solar radiation reaches the higher latitudes, and the strength of our seasonal cycles. At times, it is thought that the cycles have coincided, magnifying their effects.

Anthropocene

Humans are part of a bushy sideshoot of apes that proliferated across the globe in the Quaternary, eventually shaping the entire atmosphere and altering the land and seas entirely.

Some researchers have suggested adding a third time period to the Quaternary: the Anthropocene. This would mark the most recent few hundred to few thousand years, when humankind's impact on Earth's biosphere has become measurable not only in our effect on biological life, but also on the rock record itself. Although not officially recognized as a geological time period currently, the Anthropocene is commonly used to refer to the last few hundred years since the Industrial Revolution.

We now face what is called the Sixth Mass Extinction. The scale of disappearances among living things, and damage to the Earth itself, is such that it would undoubtedly be visible in the fossil record. We face an uncertain future as we work to combat climate change, but not only are we the first animal with the power to radically damage our planet, we are also the first with the ability to save it.

Woolly Mammoth — Ice Age Specialists

The woolly mammoth was a kind of elephant that lived on the Ice Age steppe of the northern hemisphere. It was specially adapted to thrive in persistent cold. Mammoths became extinct only around four thousand years ago, and their fossils have yielded DNA, teaching us about the genetics of ice-age survival, and putting them at the heart of the debate about bringing long-lost animals back to life.

Woolly mammoths (*Mammuthus primigenius*) are one of the most recognizable extinct animals from the Ice Age. These gargantuan herbivores lived from around forty thousand years ago until just four thousand years ago. They had long, curved tusks, high foreheads, and the northernmost species had overgrown, shaggy fur that was pale to dark chocolate brown in colour. They loomed over 3 metres (10 feet) at the shoulder, weighing as much as 6 tonnes – similar to a modern African elephant. Males and females both had impressive tusks for defence against predators that might attack their calves, as well as for competition with other mammoths for territory and mates. These behemoths of the recent past would have roamed for miles across the spartan landscape, raising generations of herds through the bitter chill of the ice ages.

Although we tend to think of them as being a single species, there were in fact lots of different species and sub-species of *Mammuthus*. DNA studies tell us that their closest living relatives are Asian elephants, but they probably originally evolved from an African species called *M. subplanifrons*, which eventually spread across all of Eurasia. Mammoths migrated into North America thanks to low sea levels, which exposed a land bridge connecting Siberia with Alaska. One of the largest mammoths was the North American steppe mammoth (*M. trigontheri*), which reached a staggering 4 metres (13 feet) in height. The smallest was the pygmy mammoth (*M. exilis*), which lived on the Channel Islands off the coast of California, and was only as tall as a human.

The cause of the woolly mammoth's extinction is much debated, but climate change and human hunting were the main factors. Their preferred habitats shrank as the climate warmed, but their decline coincided with the spread of humans, who we know from archaeological evidence hunted them. It may be that the slow reproductive rate of mammoths meant they couldn't replenish their populations

Mammoths are among the best known extinct animals, leaving their bones, tusks, and frozen bodies all across the northern hemisphere.

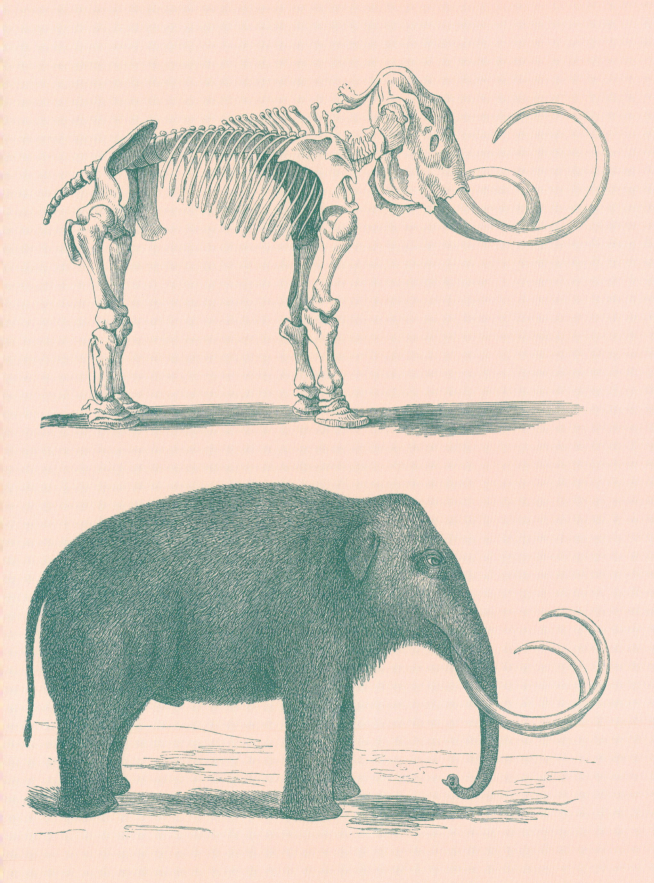

quickly enough to survive the human onslaught. The last mammoths survived on Wrangel Island in the Arctic Ocean. Cut off from the Siberian mainland by rising sea levels, they persisted until just 3,700 years ago. DNA studies reveal they suffered harmful genetic mutations from inbreeding, which affected their fat reserves and general health. Hunting by humans finally hastened their extinction.

Mega-herbivores

Mammoths thrived in the cold, dry tundra prevalent in the Pleistocene. This comprised low-growing grasses, sedges and herbs, with scattered shrubs and trees in the southernmost reaches. Mammoths shared this landscape with woolly rhinoceros, bison and horses, as well as wolves, sabre-toothed cats, cave hyenas and bears.

Mammoths munched on masses of vegetation: up to 185 kilograms (408 pounds) every day, the equivalent of ten hay bales. Their muscular trunks tore up grass and pulled leaves from branches, changing the landscape. Like modern elephants, mammoths replaced their molars repeatedly as adults, with new ones growing in from the back of the jaw while the old ones fell out at the front, as though on a conveyor belt. In modern elephants this happens six or seven times in their lifetime. Mammoths had the most complex ridged teeth of any known elephant, and large muscles in their domed skulls powered constant chewing. Frozen baby mammoths have been found with adult dung in their stomach, which they ate because their teeth weren't yet fully formed to grind fresh plants for themselves.

Cold-weather Specialists

Adaptations for cold climates are a mammal specialty. The world has rarely been as cold as it is today, and mammals have fur and are warm-blooded, priming them to withstand extremes of temperature. Alongside mammoths, other animals thrived in the ice ages, including giant ground sloths, short-faced and cave bears, giant beavers and mastodon – the latter being a type of elephant not closely related to mammoths. Some ice-age animals persist today, such as bison and musk oxen, caribou, Arctic ground squirrels, walrus and saiga antelope. Most are confined to the furthest latitudes of the planet, where the climate suits them best.

To survive the chill at the glacier's edge, woolly mammoths had thick fur with coarse shaggy guard hairs up to a metre (3 feet) long. These covered a dense short under-coat of insulating wool. Their fur stayed glossy and water-repellent with oils secreted from their skin. Mammoths probably moulted seasonally, scratching themselves on rocks to dislodge hairs. Evidence for this is found on the surface of boulders, worn smooth at mammoth shoulder height by hundreds of years of habitual rubbing. Mammoth ears were extremely small and their tails short, minimizing frostbite when the temperature plummeted. Their characteristic humpbacked appearance was created by a thick layer of stored fat above their neck and shoulders, as much as 10 centimetres (4 inches) deep. It was even present in very young calves. The fat insulated them from the elements as well as providing energy when food was scarce.

The mammoths' adaptations for the cold can even be traced in their DNA, extracted from bodies frozen for thousands of years in permafrost in the Yukon in North America, and northern Siberia. Icy conditions usually hinder the body's ability to deliver oxygen to cells, but mammoth genes reveal they had mutations in their haemoglobin to help deliver oxygen efficiently around their bodies despite the persistent cold.

Mammoths and Humans

Mammoths play an important role in human history and culture. Humans (including *Homo erectus* and Neanderthals) co-existed with them, hunting them for meat and using their bones, hides and ivory for tools, clothing and carved objects. Mammoths are the third-most depicted animal in prehistoric rock art after bison and horses. These artworks show us that mammoths lived in herds. In eastern Europe and Russia between 15,000 and 40,000 years ago, people used mammoth bones to build shelters. Indigenous Siberians often carved mammoth ivory they found in the permafrost, and traded it as far as China and western Europe. The bones of mammoths were once thought to belong to spirits or large underground animals.

In the 18th and 19th centuries, Western scientists believed mammoths and their relatives, mastodons, might still exist in parts of the New World not yet reached by Europeans. In the early 1800s, explorers Meriwether Lewis and William Clark were asked by United States President Thomas Jefferson to look for these animals during their travels in North America. They found none alive, but brought back bones that helped scientists study extinct elephants, contributing to a growing acceptance of the concept of extinction.

Dodo — Unnatural Extinctions

The dodo disappeared so swiftly after its first encounter with humans that by the time most of the world had heard of this wonderful bird, it was already lost forever. Found only on Mauritius, the dodo has become the symbol of the most unnatural extinction event in Earth's history. Despite calls for animals lost under our heavy human footprint to be cloned back to life, the accelerating biodiversity crisis will require more than a quick de-extinction to fix. Otherwise we may all go the way of the dodo.

With its rotund body and large head and beak, the dodo (*Raphus cucullatus*) is a ludicrous-looking poster child for extinction. Despite our historically comical image of the dodo, in truth we know surprisingly little about this flightless bird, despite having driven it to extinction only 360 years ago. Its remains reveal it stood about a metre (3 feet) in height, but the only original descriptions of the bird made by visitors to its island home of Mauritius vary a great deal. Some say it was grey, some brown, with sleek feathers or fluffy and messy, sometimes with a colourful beak, other times not. This confusion may be due to seasonal changes as it moulted, or differences between male and female birds. Many of the most famous dodo images were created long after their extinction. Other than scattered bones, the only soft tissue that remains is a single dried head in the Oxford University Museum of Natural History in England.

The clumsy, stupid image of this animal is a fabrication; in truth, it was well-adapted for its ecosystem. The dodo evolved from flying ancestors that arrived on an isolated island free of ground-based predators. With abundant food supplies, they lost the need and ability to fly, becoming entirely terrestrial. They probably ate fallen fruit, nuts, seeds and roots, and ingested small stones called gastroliths to aid digestion – a trick first used by their ancient dinosaur relatives. The closest living relative of the dodo is the Nicobar pigeon, a flamboyant species from India and the Malay Archipelago. It was also related to the Rodrigues solitaire (*Pezophaps solitaria*), which lived on a neighbouring island and was likewise extinguished by the arrival of humans. Their disappearance is no surprise when you read accounts such as that made by the crew of the Dutch ship *Bruin-Vis*. In one single visit in 1602, they killed 25 dodos, providing more food than they could eat in one meal.

The dodo has become a shorthand for extinction. It was hunted to oblivion in less than a human lifetime.

Like many animals in locations free of land-based predators, the dodo was fearless of humans. This made it an easy target for hungry sailors and the animals they brought with them, such as dogs and pigs, which raided their nests. Just 64 years after the first record of this wondrous bird, it was gone. This alarming loss helped people realize for the first time that humans could eradicate entire species forever. The dodo is the symbol of extinction. Sadly, it is far from the last of our planet's incredible animals to be snuffed out of existence by the carelessness of humankind.

Natural Loss

Extinction occurs when the last organism belonging to a species dies. Species can become 'functionally extinct' long before this, if there are too few remaining animals to replenish the population. Although often a negative event, extinction is also a natural part of life on Earth. New species evolve through the process of speciation, and extinction is the flip side of that coin. On average, a species exists for a few million years before being replaced by other organisms or evolving into new forms. Almost all species that have ever lived are now extinct, amounting to billions of unique organisms. Despite the staggering diversity of life on Earth today, it is a tiny slice of what we see in the epic reach of the planet's fossil record.

Mass extinctions are relatively rare, occurring when there is a sharp decline in biodiversity and the rate of extinction outstrips the rate of speciation. There have been five major mass extinctions in the last 550 million years, and many smaller ones. The size of a mass extinction is measured by the percentage of species or genera that become extinct, rather than the number of actual organisms. For example, if millions of animals belonging to a single species died out, it would constitute an extinction, whereas if only a few thousand animals died, but they included hundreds of different species, that would be a larger, mass extinction event. This is because the disappearance of multiple species at the same time has a far greater impact on food webs and ecosystems, with long-term consequences for life on Earth.

Unlike previous extinctions caused by natural disasters such as volcanic eruptions, our current, sixth mass extinction has been caused by humans. We know from the fossil record that the speed of the current extinction outstrips previous events, and it shows no sign of slowing. Some scientists argue that it began long ago, when

The only surviving soft tissue of the dodo — a head with skin attached (top) — can only hint at the original appearance of this magnificent animal (interpreted reconstruction, bottom).

232

humans began their march around the globe. Between 132,000 and 1,000 years ago, around 177 large mammals became extinct, and studies suggest that the spread of humans accounts for at least 64 per cent of those extinctions. This means we have had a greater impact on biodiversity than natural changes in climate, habitat or even the end of the last ice age.

The International Union for Conservation of Nature (IUCN), a global inventory of our planet's biodiversity, found that 27 per cent of the species it surveyed are currently at risk of extinction. At least 571 species have disappeared in the last three centuries, and as many as one million plants and animals are now in danger of being lost forever.

Hunting and fishing comprise part of the threat, but habitat loss through human development for farming and building is perhaps the largest overall cause of extinction. Apart from the tragic loss of biodiversity that evolution has taken 550 million years to accrue, these extinctions threaten the future of humankind. The decimation is dismantling the balance of entire nutrient cycles and food webs. The loss of resources and ecosystems services such as cleaning water and drawing down carbon have profoundly negative impacts both on the planet and on ourselves.

Bringing Them Back

In the last few decades there has been mounting interest in the idea of resurrecting extinct species, or 'de-extinction'. Although previously the realm of fantasy, new technologies such as DNA extraction and cloning have made the idea appear tantalizingly within reach. Apart from the scientific difficulties in making the dream a reality, bringing lost species back from the dead raises ethical concerns that researchers and the public alike must grapple with.

One of the more achievable options for de-extinction is selectively breeding animals for traits found in their extinct ancestors. This form of artificial selection is straightforward, but whether the results truly represent the lost animal is debatable. More controversially, it has been proposed that extinct animals could be cloned. This requires well-preserved remains, and it has been suggested for the woolly mammoth. The mummified frozen bodies of mammoths are often preserved with pelts of hair, skin and organs, which researchers have used to map their genes and proteins. Some people believe they could use this DNA to bring mammoths back, mixing it with that of living elephants to fill in the genetic gaps.

De-extinction, as well as being technically challenging – perhaps even impossible, does not solve the root causes of our ongoing extinction crisis. If such animals somehow survived the fragile process of cloning, it is unclear what the benefit would be now that the climate and ecosystems they inhabited have disappeared. Critics point out that the enormous expense and effort would be better focused on conservation and developing new, sustainable technologies to halt the destruction of our natural world. Until we address habitat loss, pollution and climate change, it is probably unwise to resurrect the many animals we've extinguished. Whatever happens, with so few physical remains of the wonderful and singular dodo, it is unlikely that this bird will ever walk the Earth again.

Drosophila — Scientific Insects

Flies have shaped our modern world — and our future. They indirectly kill more people each year than any other creature, but are foundational to our knowledge of evolution, biology and medicine. *Drosophila* is a fly of science, pervading laboratories around the world — it has even been to space. This humble creature teaches us about the workings of our world and ourselves.

Drosophila melanogaster may not be a familiar name to everyone, but this tiny fly is found on every continent. Also known as a fruit fly, the adults are only around a millimetre in length, pale yellow to brown or black, with red eyes. It originally came from Africa, and is not to be confused with other 'fruit flies' that are agricultural pests and health hazards. Annoying as it may be in your kitchen, *Drosophila* transmits no diseases and is essentially harmless. This miniscule creature is one of the most important animals in human research, and the source of six Nobel Prizes in science.

Drosophila began their careers in science in the 1900s. They reproduce quickly, are easy to keep, have a small genome, and their use in research experiments raises few ethical objections. This makes them ideal for the study of heredity: the way biological traits are passed from parents to offspring. Many breakthroughs in our knowledge of genetics and the role of environment in evolution come from studying fruit flies, and they continue to be used in countless ways as a 'model organism'. Despite the millennia separating our branches on the tree of life, humans still share around 60 per cent of genes with *Drosophila*, making them useful in medical studies, including research for cancer treatments and combatting Alzheimer's disease.

The word 'fly' barely captures the staggering breadth of this group's evolutionary diversity. Although many insects are known as 'flies', technically only those in the order Diptera are flies. The first fossil dipterans come from the Triassic, and today they include craneflies, gnats, horse flies, hoverflies, midges and mosquitos. These insects have a single pair of wings, making them highly manoeuvrable; large compound eyes; and piercing and sucking mouthparts for feeding. Each of their six legs ends in tiny claws with 'pads' that produce an electrostatic force, allowing them to cling to even the smoothest surface.

We owe many scientific advances to *Drosophila melanogaster*, the fruit fly. These tiny 'model organisms' have even been launched into space.

Today, flies are found globally except in Antarctica, and although over 150,000 species have been described, there are many more still to be discovered. They are second only to bees as global pollinators, and may have been among the first to provide that service. The largest fly is *Gauromydas heros*, which can grow longer than a human finger, with a wingspan of 10 centimetres (4 inches). Most flies are much smaller, one of the tiniest being *Euryplatea*, which parasitizes equally diminutive ant species by laying eggs in their heads, eating them from the inside out. Many flies are parasitic, laying eggs in the flesh of other animals. Some feed on carrion, dung, plant matter or fungi, making them important recyclers in ecosystems around the globe. Flies locate food by detecting scent and taste on air currents, and also using chemoreceptors in their feet, allowing them to taste food by walking across it. Blood-sucking insects detect the carbon dioxide breathed out by animals, or sense body heat to pinpoint their target. Flies are also an enormously abundant source of food for other insects, animals and even plants, such as the Venus flytrap (*Dionaea muscipula*).

Lord of the Flies

There is no doubt that flies have a dark side. In most cultures they are considered a source of disease and even evil, despite the many important positive contributions they make to food webs. They are rightly associated with illness and decay: many infectious diseases are transmitted by fly bites – notably from mosquitos. Houseflies spread food-borne diseases throughout the world, enjoying the warmth of human habitations and our glut of food to land on. Flies also spread diseases among livestock, such as screwworm, and they can damage crops.

Mosquitos alone are the biggest killers of humans on Earth today because they spread dengue fever, West Nile and yellow fever, Zika virus and malaria, among others. As vectors of disease, they account for well over a million deaths every year. This deadly spread has even been used as a weapon in warfare: in the Second World War, low-flying airplanes dropped bombs packed with flies and a slurry of cholera onto parts of China, killing at least 440,000 people.

In many cultures flies are associated with mortality, for example as symbols of Nergal, the Old Babylonian god of death. In Christian theology, the devil demon Beelzebub is known as the Lord of the Flies. They feature in art, from fly-shaped lapis lazuli beads in ancient

Mesopotamia, to depictions in surrealist paintings. For all the horror they may instil, theirs is a pervasive company we cannot escape.

Model Organism

Fruit flies have the distinction of being the first animals we sent into space. In 1947 the V-2 rocket carried fruit flies and moss out of our atmosphere, before they returned to Earth in a special capsule with its own parachute. The intention was to study the effect of radiation exposure, and the fruit flies returned showing no signs of mutation caused by radiation. *Drosophila* continues to accompany astronauts on manned shuttle missions, spending extended time in space. This allows researchers to study the effects of space travel on their immune system and biology, and develop ways to keep astronauts safe and healthy – an important part of preparations for longer manned flights to the moon and Mars scheduled in the coming decades.

Drosophila are not the only flies that have proven useful to science. At crime scenes they are always first to arrive, and the presence of their larvae on corpses allows forensic scientists to make surprisingly accurate estimates for time of death and how bodies were disposed. One of the flies that uses corpses in this way are blow flies and their larvae, or maggots. Maggots can also be used in medicine to clean dead tissue from wounds.

Famously, in the fiction novel *Jurassic Park*, blood-sucking mosquitos preserved in amber provide a source for dinosaur DNA, used to clone these creatures and bring them back to life. This, however, is pure fiction: cloning has not yet reached the point where de-extinction is a reality even for recently lost organisms. Even if it had, DNA can't survive for millions of years, and if it did, the stomach of biting flies is an unlikely source of intact genetic material.

Homo – Human Evolution

What makes the human animal unique? In the geological whisker-twitch that we have inhabited this planet, we have transformed it radically. Humans are creatures of the Quaternary, forged by the anvil strikes of ice ages and climate change. We've used nature's incredible resources to build our lives, satisfying hunger, constructing shelters, creating art and myths. From the earliest stone technologies to orbiting satellites, our journey is extraordinary, but our destructive presence will also tarnish the rock record for millennia to come.

We are the last of many human species to evolve over the past few million years. Our modern technologies and civilizations are only thousands of years in the making – dwarfed by the grandeur of geological time.

Humans (*Homo sapiens*) tend to place themselves separate from other animals. But we are just another primate, sharing around 99 per cent of our DNA with chimpanzees and bonobos. Modern humans have existed for approximately 150,000 years. We are relatively hairless creatures, with large, expanded brains. Our extensive grey matter has enabled the development of complex technology, language, self-reflection, art and music. Despite this, we have often been the victims of the animals we live alongside; fossils of early *Homo* species include those eaten by crocodiles and leopards. Deaths by large wild predators still occur, although they are far outweighed by the tens of thousands of deaths by dog attacks, snakes and mosquitos. Despite being born helpless, ageing slowly and lacking sharp teeth, claws or the ability to run fast, it is our sophisticated cognition that has equipped us for survival, making us a formidable species – and the architects of the sixth mass extinction on our planet.

Primate origins can be traced back to small spindly creatures such as *Plesiadapis,* which lived just after the end-Cretaceous mass extinction. Between 15 and 20 million years ago, the last common ancestor of all the living apes emerged, and 8 million years ago lived our last common ancestor with gorillas. Our lineage and that of chimpanzees split more recently, between 7 and 4 million years ago. The oldest fossils from our own genus, *Homo*, appear a mere 2.3 million years ago, near the start of the Quaternary, making this time period the true age of humankind.

We used to imagine human evolution as a straight line of descent, but more than 15 species of early humans, or hominins, have existed in the last few million years. They often co-existed, even interbreeding. Fossil discoveries in the last fifty years have added many new faces to the story of our evolution, but there is fierce debate over where they sit in our family tree. What is certain is that the term

'human' includes greater diversity than we might have previously imagined. From scattered bands of people harnessing fire and creating tools, there are now 7.8 billion humans and counting, most of us living in cities. Since the Industrial Revolution, humans have accelerated their impact on the Earth. The very intelligence that makes us so unique also threatens the survival of the planet as we know it.

What Makes Us Human

Humans are distinguished particularly by our upright, bipedal stance, our use of language and the ability to make tools – although all of these abilities are present in other animal groups. We have large brains compared to our body mass – nearly three times the size of a chimpanzee or gorilla. Some researchers suggest that as climate and habitat changed in Africa, we developed bipedalism to save energy and cover long distances. This incidentally freed our hands, which found use foraging and carrying. Walking on two legs has had radical effects on our skeleton, re-orienting and altering the shape of the pelvis, spine and leg joints. Although advantageous in some ways, this had many drawbacks: humans suffer back and joint problems associated with our upright posture, and have much more difficult births than other primates. The balance between brain size and the size and orientation of the birth canal probably contributes to how helpless and undeveloped human babies are at birth. Whereas most other primates can walk, and cling to and mimic adults, almost immediately after birth, humans have a protracted childhood and comparatively delayed sexual maturity.

We can track increased brain size through the different species of human in the fossil record, but probably the structure of the brain was as important as its size. We rely on archaeology to find clues for when complex culture emerged. Technology in the form of stone tools (lithics) provides evidence, and researchers study how these were produced to track development in different human species. The first identifiable stone-working dates to around 3.3 million years ago, probably from the hominin *Australopithecus*. By the start of the Quaternary, the oldest member of our genus, *Homo habilis*, had developed new stone technologies for butchering animals and processing hides, plants and wood.

Another thread of evidence for our mental and social development comes from the way early hominins dealt with their dead. Some of the oldest intentional human burials are known from Israel, where ten bodies of *Homo sapiens* were placed carefully in a cave named Mughâret-es-Skūl. These date to around 100,000 years old, telling us that by that point at least, ancient humans had complex social practices surrounding death. Other material culture includes collaborative hunting and skin-working, skills shared by Neanderthals and early *Homo sapiens* at least 200,000 years ago. Representational art such as cave paintings and carved objects are considered important signs of fully modern human cognition, and only appear fifty thousand years ago.

Dispersal and Impact

The earliest origins of *Homo sapiens* lie within Africa. Some of those humans began moving into Eurasia earlier, but ancient DNA tells us that all living humans come from a small population that existed around 100,000 years ago. Dispersals probably took place multiple times, making the picture of our emergence hard to interpret. We know these humans encountered other hominin species, including *Homo neanderthalensis* (Neanderthals) and the Denisovans. Interbreeding between them seems to have been common, and up to 6 per cent of modern human DNA can be traced to this mixing. Today, humans are the only surviving hominins. We reached East Asia at least seventy thousand years ago, finding our way across Indonesia to Australia a few millennia later. Humans are not known in central or western Europe until 42,000 years ago. Others may have made the crossing to the Americas around this time – a journey repeated at the tail end of the last ice age.

Humankind has had an enormous impact on animal diversity in the last 100,000 years. The appearance of humans in continents around the world often coincided with the disappearance of multiple species of animals, particularly big mammals that were hunted for game. Not only did humans affect animal populations, we have also shaped habitats. Since the end of the last ice age, we have had an impact as hunter-gatherers, pastoralists and eventually farmers. Plant selection, deforestation and changes in fire regimes are evident even in environments many consider pristine and untouched, such as the Amazon rainforest. In particular, the use of fire on grasslands to increase productivity of the land has radically changed ecosystems, as well as altering water cycles and releasing carbon dioxide. As our populations have increased, these impacts have grown, and the recent Industrial Revolution rapidly accelerated the adverse effects of humans on our planet in the last two hundred years.

Our Future

Humans are undoubtedly one of Earth's most extraordinary inhabitants. However, many of the things we consider hallmarks of our species are not so unique in the context of all animal kind. Other creatures have much larger brains (for example, elephants), use tools (crows), communicate in complex social groups (dolphins), farm (ants) and have global impacts on climate and ecosystems (algae and earthworms). It is true to say that in humans, however, all of these traits combine. Coupled with our technologies, large populations and social networks, we achieve things never seen before on this planet. Adapting to solve the problems of climate change and increasingly scarce resources are two of the major challenges we face. Scientists agree climate change is occurring, but how severe it will be depends on the efforts made to reduce greenhouse gas emissions.

In the most distant future, whether humans curb their negative impact on Earth or disappear from it entirely, the planet will right itself naturally. Our existence will be sandwiched between the rock layers, a thin layer of destruction, like the asteroid impacts of the past. Based on past extinctions, it is likely to take around ten million years for life to recover from the current human-induced mass extinction. A new supercontinent will coalesce over the next 250 million years, transforming the global climate all over again. Our nearly exhausted fossil fuel reserves will have been replenished by that time, with millennia upon millennia of uninterrupted natural growth, decay and burial.

After 500 million years, the increasing heat of the sun will start to disrupt the delicate Goldilocks balance of Earth's cycles. By one billion years in the future, complex life on this planet is unlikely to survive. Although geological time appears vast to us, in the grand scheme of the universe, all complex organisms on Earth have evolved in a short cosmic in-breath. Who knows what wonders natural selection has produced on the many other habitable worlds glinting briefly into life elsewhere in the universe?

PICTURE CREDITS

ACKNOWLEDGEMENTS

Thank you for reading this book, I hope you enjoyed it. Picking only a few organisms to write about, from a planet bursting with millions of mind-boggling forms alive and extinct, was extremely difficult. This is by no means a final selection, but I hope it covers a few of the star players that have peppered every part of this world over the last 4.6 billion years. I've loved immersing myself in the science of organisms as different as earthworms and kelps – every one of them deserves a book of its own!

I owe a debt of gratitude to the wonderful fellow experts who read parts of this book that pertained to their scientific specialisms, to fact-check and give feedback: Gwen Antell, Jordan Bestwick, Neil Brocklehurst, Mark Carnall, Albert Chen, Richard Dearden, Paige dePolo, Frankie Dunn, Ricardo Perez-De-La Fuente, Davide Foffa, Russell Garwood, Sandy Hetherington, Femke Holwerda, Susannah Lydon, Imran Rahman, Paul Smith, Christine Strullu-Derrien, and Becky Wragg Sykes all took time to do this, and I am so grateful. Let me know if I can return the favour! I'm sorry that not all of the chapters reviewed made it to the final cut, but know that your input was vital to the process.

A huge thank you goes to Kerry Enzor for taking me onboard to write this, and Julia Shone and Will Webb for creating something incredibly beautiful from all those words and ideas. I was delighted that emerging artist and palaeontologist Grace Varnham contributed illustrations, as well the artists recent and historical whose work is featured in these pages. Thanks also to all the unnamed folk who have been involved in producing the book, from proof reading to printing.

I'm grateful to my colleagues at the Oxford University Museum of Natural History, especially Paul Smith, for their support and enthusiasm about this book. It is even inspiring a new exhibition in the galleries – I'm honoured to provide the starting seed.

Nothing I write would happen without the patient support of my partner Matt, who always sees the value in everything I try to achieve, and makes sure I have the space, time, and hot dinners necessary to fuel my work. Thank you for all the tea, love and thousands of conversations. Finally, thanks to Mjölnir, who was no help at all in the writing process, but reminded me at least three times a day to take a break from working. It may have only been in order to provide her with biscuits, but it is an important role nonetheless.

SELECTED REFERENCES AND FURTHER READING

Anderson, P.S. and Westneat, M.W., 2007, 'Feeding mechanics and bite force modelling of the skull of Dunkleosteus terrelli, an ancient apex predator', *Biology Letters*, 3(1), pp.77–80.

Anderson, F.E., Williams, B.W., Horn, K.M., Erséus, C., Halanych, K.M., Santos, S.R. and James, S.W., 2017, 'Phylogenomic analyses of Crassiclitellata support major Northern and Southern Hemisphere clades and a Pangaean origin for earthworms', *BMC Evolutionary Biology*, 17(1), pp.1–18.

Arratia, G., 2013, 'Morphology, taxonomy, and phylogeny of Triassic pholidophorid fishes (Actinopterygii, Teleostei)', *Journal of Vertebrate Paleontology*, 33 (sup1): 1–138.

Benton, M.J. and Harper, D.A., 2020, *Introduction to Paleobiology and the Fossil Record*, John Wiley & Sons.

Benton, M. J., 2021, *Dinosaurs: New Visions of a Lost World*, Thames and Hudson Ltd.

Betts, Holly C.; Puttick, Mark N.; Clark, James W.; Williams, Tom A.; Donoghue, Philip C.J.; Pisani, Davide, 2018, 'Integrated genomic and fossil evidence illuminates life's early evolution and eukaryote origin', *Nature Ecology & Evolution*, 2 (10): 1556–1562.

Bindeman, I.N., Zakharov, D.O., Palandri, J., Greber, N.D., Dauphas, N., Retallack, G.J., Hofmann, A., Lackey, J.S. and Bekker, A., 2018, 'Rapid emergence of subaerial landmasses and onset of a modern hydrologic cycle 2.5 billion years ago', *Nature*, 557(7706), pp.545–548.

Black, Riley, 2021, *Last Days of the Dinosaurs*. St. Martin's Publishing Group.

Briggs, D.E., Clarkson, E.N. and Aldridge, R.J., 1983, 'The conodont animal', *Lethaia*, 16(1), pp.1–14.

Brocks, J.J., Jarrett, A.J., Sirantoine, E., Hallmann, C., Hoshino, Y. and Liyanage, T., 2017, 'The rise of algae in Cryogenian oceans and the emergence of animals', *Nature*, 548(7669), pp.578–581.

Clack, J. A., 2009, 'The fin to limb transition: new data, interpretations, and hypotheses from paleontology and developmental biology', *Annual Review of Earth and Planetary Sciences*, 37: 163–179.

Danforth, B.N. and Poinar, G.O., 2011, 'Morphology, classification, and antiquity of Melittosphex burmensis (Apoidea: Melittosphecidae) and implications for early bee evolution', *Journal of Paleontology*, 85(5), pp.882–891.

Donoghue, P.C. and Purnell, M.A., 2005, 'Genome duplication, extinction and vertebrate evolution', *Trends in Ecology & Evolution*, 20(6), pp.312–319.

Dunn, F.S., Liu, A.G., Grazhdankin, D.V., Vixseboxse, P., Flannery-Sutherland, J., Green, E., Harris, S., Wilby, P.R. and Donoghue, P.C., 2021, 'The developmental biology of Charnia and the eumetazoan affinity of the Ediacaran rangeomorphs', *Science Advances*, 7(30).

Field, D.J., Benito, J., Chen, A., Jagt, J.M.W., Ksepka, D.T., 2020, 'Late Cretaceous neornithine from Europe illuminates the origins of crown birds', *Nature* 579 397–401.

Ford, D.P. and Benson R.B.J., 2020, 'The phylogeny of early amniotes and the affinities of Parareptilia and Varanopidae', *Nature Ecology & Evolution* 4 (1), 57–65

Fortey, Richard, 2000, *Trilobite: Eyewitness to Evolution*, Alfred a Knopf Inc.

Fraser, N.C. and Suess, H.D., 2017, *Terrestrial Conservation Lagerstätten: Windows into the Evolution of Life on Land*, Dunedin Academic Press.

Fritsche, Olaf, Foitzik, Susanne, 2021, *Empire of Ants: The Hidden Worlds and Extraordinary Lives of Earth's Tiny Conquerors*, Gaia.

Garwood, Russell J., 2012, 'Patterns In Palaeontology: The first 3 billion years of evolution', *Palaeontology Online*, 2 (11): 1–14.

Gemmell, N.J., Rutherford, K., Prost, S., Tollis, M., Winter, D., Macey, J.R., Adelson, D.L., Suh, A., Bertozzi, T., Grau, J.H. and Organ, C., 2020, 'The tuatara genome reveals ancient features of amniote evolution', *Nature*, 584(7821), pp.403–409.

Gordon, Helen, 2020, *Notes from Deep Time: A Journey Through Our Past and Future Worlds*, Profile Books.

Graham, J.B., Aguilar, N.M., Dudley, R. and Gans, C., 1995, 'Implications of the late Palaeozoic oxygen pulse for physiology and evolution', *Nature*, 375(6527), pp.117–120.

Hallam, Tony, 2005, *Catastrophes and Lesser Calamities: The Causes of Mass Extinctions*, Oxford University Press.

Hetherington A.J., 2019, 'Evolution of plant rooting systems', in *eLS*, John Wiley & Sons, Ltd, Chichester.

Hu; et al., 2005, 'Large Mesozoic mammals fed on young dinosaurs', *Nature*, 433 (7022): 149–152.

Hunt, T., Bergsten, J., Levkanicova, Z., Papadopoulou, A., John, O.S., Wild, R., Hammond, P.M., Ahrens, D., Balke, M., Caterino, M.S. and Gómez-Zurita, J., 2007, 'A comprehensive phylogeny of beetles reveals the evolutionary origins of a superradiation', *Science*, 318(5858), pp.1913–1916.

Hume, J.P., 201, 'The Dodo: from extinction to the fossil record', *Geology Today*, 28(4), pp.147–151.

Janis, K.M., Scott, KM, and Jacobs, LL., 1998, *Evolution of Tertiary Mammals of North America: Volume 1, terrestrial carnivores, ungulates, and ungulate like mammals*, Cambridge University Press.

Jeram, A.J., Selden, P.A. & Edwards, D., 1990, 'Land animals in the Silurian: arachnids and myriapods from Shropshire, England', *Science* 250: 658–666.

Kemp, T.S., 2005, *The Origin and Evolution of Mammals*, Oxford University Press.

Kluessendorf, J. and Doyle, P., 2000, 'Pohlsepia mazonensis, an early "octopus" from the Carboniferous of Illinois, USA', *Palaeontology*, 43(5), pp.919–926.

Larson, G., Piperno, D.R., Allaby, R.G., Purugganan, M.D., Andersson, L., Arroyo-Kalin, M., Barton, L., Vigueira, C.C., Denham, T., Dobney, K. and Doust, A.N., 2014, 'Current perspectives and the future of domestication studies', *Proceedings of the National Academy of Sciences*, 111(17), pp.6139–6146.

Lawrence, N., 2015, 'Assembling the dodo in early modern natural history', *The British Journal for the History of Science*, 48(3), pp.387–408.

Long, J.A., Mark-Kurik, E., Johanson, Z., Lee, M.S., Young, G.C., Min, Z., Ahlberg, P.E., Newman, M., Jones, R., den Blaauwen, J. and Choo, B., 2015, 'Copulation in antiarch placoderms and the origin of gnathostome internal fertilization', *Nature*, *517*(7533), pp.196–199.

Luo, Z.X., 2007, 'Transformation and diversification in early mammal evolution', *Nature*, 450(7172), pp.1011–1019.

Maletz, J., 2017, *Graptolite Paleobiology*, John Wiley & Sons.

Marjanovic, D. and Laurin, M., 2013, 'The origin(s) of extant amphibians: a review with emphasis on the "lepospondyl hypothesis"', *Geodiversitas*, 35(1), pp.207–272.

Martinetto, Edoardo, Tschopp, Emanuel and Gastaldo, Robert A., 2020, *Nature through Time*, Springer Verlag.

Mayor, A., 2011, *The First Fossil Hunters*, Princeton University Press.

Miller, K.G., Kominz, M.A., Browning, J.V., Wright, J.D., Mountain, G.S., Katz, M.E., Sugarman, P.J., Cramer, B.S., Christie-Blick, N. and Pekar, S.F., 2005, 'The Phanerozoic record of global sea-level change', *Science*, 310(5752), pp.1293–1298.

Mitchell, R.L., Cuadros, J., Duckett, J.G., Pressel, S., Mavris, C., Sykes, D., Najorka, J., Edgecombe, G.D. and Kenrick, P., 2016, 'Mineral weathering and soil development in the earliest land plant ecosystems', *Geology*, 44(12), pp.1007–1010.

Motani, R., 2005, 'Evolution of fish-shaped reptiles (Reptilia: Ichthyopterygia) in their physical environments and constraints', *Annual Review of Earth and Planetary Sciences*, 33: pp. 395–420.

Neild, Ted, 2008, *Supercontinent: Ten Billion Years in the Life of our Planet*, Granta Books.

North West Highlands Geopark, www.nwhgeopark.com

Outram, A.K., Stear, N.A., Bendrey, R., Olsen, S., Kasparov, A., Zaibert, V., Thorpe, N. and Evershed, R.P., 2009, 'The earliest horse harnessing and milking', *Science*, *323*(5919), pp.1332–1335.

Panciroli E., 2021, *Beasts Before Us: The Untold Story of the Origin and Evolution of Mammals*, Bloomsbury Sigma.

Piperno, D.R. and Sues, H.D., 2005. Dinosaurs dined on grass. Science, 310(5751), pp.1126-1128.

Prasad, V., Strömberg, C.A., Alimohammadian, H. and Sahni, A., 2005, 'Dinosaur coprolites and the early evolution of grasses and grazers', *Science*, 310(5751), pp.1177–1180.

Retallack, G.J. and Landing, E., 2014, 'Affinities and architecture of Devonian trunks of *Prototaxites loganii*', *Mycologia*, 106(6), pp.1143–1158.

Shapiro, Beth, 201, *How to Clone a Mammoth: The Science of De-Extinction*, Princeton University Press.

Sheldrake, Merlin, 2020, *Entangled Life: How Fungi Make Our Worlds, Change Our Minds and Shape Our Futures*, Bodley Head.

Shikama, T., Kamei, T. and Murata, M., 'Early Triassic Ichthyosaurus, Utatsusaurus hataii Gen. et Sp. Nov., from the Kitakami Massif, Northeast Japan', *Science Reports of the Tohoku University Second Series (Geology)*, 1977. 48(1–2): p. 77–97.

Shu, D.G., Luo, H.L., Morris, S.C., Zhang, X.L., Hu, S.X., Chen, L., Han, J.I.A.N., Zhu, M., Li, Y. and Chen, L.Z., 1999, 'Lower Cambrian vertebrates from south China', *Nature*, 402(6757), pp.42–46.

Slack, K.E., Jones, C.M., Ando, T., Harrison, G.L., Fordyce, R.E., Arnason, U. and Penny, D., 2006, 'Early penguin fossils, plus mitochondrial genomes, calibrate avian evolution', *Molecular Biology and Evolution*, 23(6), pp.1144–1155.

Sohn, J.C., Labandeira, C.C. and Davis, D.R., 2015, 'The fossil record and taphonomy of butterflies and moths (Insecta, Lepidoptera): implications for evolutionary diversity and divergence-time estimates', *BMC Evolutionary Biology*, *15*(1), pp.1–15.

Starko, S., Gomez, M.S., Darby, H., Demes, K.W., Kawai, H., Yotsukura, N., Lindstrom, S.C., Keeling, P.J., Graham, S.W. and Martone, P.T., 2019, 'A comprehensive kelp phylogeny sheds light on the evolution of an ecosystem', *Molecular Phylogenetics and Evolution*, 136, pp.138–150.

Stephens, L., Fuller, D., Boivin, N., Rick, T., Gauthier, N., Kay, A., Marwick, B., Armstrong, C.G., Barton, C.M., Denham, T. and Douglass, K., 2019, 'Archaeological assessment reveals Earth's early transformation through land use', *Science*, 365(6456), pp.897–902.

Stewart, Amy, 2004, *The Earth Moved: On the Remarkable Achievements of Earthworms*, Algonquin Books.

Stork, N.E., McBroom, J., Gely, C. and Hamilton, A.J., 2015, 'New approaches narrow global species estimates for beetles, insects, and terrestrial arthropods', *Proceedings of the National Academy of Sciences*, 112(24), pp.7519–7523.

Strömberg, C.A., 2011, 'Evolution of grasses and grassland ecosystems', *Annual Review of Earth and Planetary Sciences*, 39, pp.517-544.

Sues, Hans-Dieter, 2019, *The Rise of Reptiles: 320 Million Years of Evolution*, Johns Hopkins University Press.

Sun, G., Ji, Q., Dilcher, D.L., Zheng, S., Nixon, K.C. and Wang, X., 2002, 'Archaefructaceae, a new basal angiosperm family', *Science*, 296(5569), pp.899–904.

Taylor, T. N., Hass, H., Remy, W. and Kerp, H., 1995, 'The oldest fossil lichen', *Nature*, 378: 244.

Taylor, P.D., 2016, 'Fossil folklore: ammonites', D*eposits Magazine*, 46, pp.20–23.

Tetlie, O.E., 2007, 'Distribution and dispersal history of Eurypterida (Chelicerata)', *Palaeogeography, Palaeoclimatology, Palaeoecology*, 252(3–4), pp.557–574.

Vickers-Rich, P. and Komarower, P. eds., 2007, *The Rise and Fall of the Ediacaran Biota*, Geological Society of London.

Wahlberg, N., Wheat, C.W. and Peña, C., 2013, 'Timing and patterns in the taxonomic diversification of Lepidoptera (butterflies and moths)', *PLOS one*, *8*(11).

Willis, K. and McElwain, J., 2014, *The Evolution of Plants*, Oxford University Press.

Wilson, H. M. & Anderson, L. I., 2004, 'Morphology and Taxonomy of Paleozoic millipedes (Diplopoda: Chilognatha: Archipolypoda) from Scotland', *Journal of Paleontology* 78: 169–184.

Witton, Mark, 2013, *Pterosaurs: Natural History, Evolution, Anatomy*, Princeton University Press.

Whalley, P.E.S., 1985, 'The systematics and palaeogeography of the Lower Jurassic insects of Dorset', *Bulletin of the British Museum* (Natural History), Geology, 39: 107–189.

INDEX

Page numbers in *italics* indicate
illustration captions.

DR ELSA PANCIROLI is a Scottish palaeontologist and writer fascinated by the natural history of our planet. She is currently a researcher at the Oxford University Museum of Natural History, and associate researcher at National Museums Scotland. She gives regular public talks for all ages, and has written about science for *The Guardian* and *BBC Science Focus*, as well as appearing on podcasts, radio and television. She is also the author of *Beasts Before Us: The Untold Story of Mammal Origins and Evolution.*

First published in Great Britain in 2022 by

Greenfinch
An imprint of Quercus Editions Ltd
Carmelite House
50 Victoria Embankment
London EC4Y 0DZ

An Hachette UK company

A CIP catalogue record for this book is available from the British Library

HB ISBN 978-1-52941-398-4
ebook ISBN 978-1-52941-399-1

10 9 8 7 6 5 4 3 2 1

Design by willwebb.co.uk
Printed and bound in Singapore by 1010 Printing

MIX
Paper from
responsible sources
FSC
www.fsc.org   FSC® C016973